Astronomy For Astro Navigation
Colour Version

Written by Lt. Cdr Jack Case, M.A. B.Ed.(Hons.)
Published by Bookcase Learning Resources
www.astronavigationdemystified.com
First Edition April 2015
ISBN-13: 9781511522083
ISBN-10: 1511522089

Books in the Astro Navigation Demystified Family:
Astro Navigation Demystified
Applying Mathematics To Astro Navigation
Astronomy For Astro Navigation
Celestial Navigation - Theory and Practice

The Astro Navigation Demystified website provides a free
resource for all those interested in the subject:
www.astronavigationdemystified.com

Preface

Astronomy is a vast, complex and very interesting subject; however, when I was taught astro navigation, I often wished that my tutor would 'cut to the chase' and focus on only those aspects of astronomy that were relevant. When I taught navigation myself, I searched high and low for an astronomy book that would allow me to pick out just those topics that I needed to teach my students but I was never able to find one. Throughout my career, I meant to write such a book myself to help other astro navigation tutors and students but I just never found the time. Now that I am retired at last, I have had the time to produce the book and I offer it, not just for navigation students but also for practicing navigators. I hope also that teachers in schools and colleges might find it a helpful tool for introducing students to astronomy as well as certain aspects of geography and mathematics.

About The Author

Jack Case is an experienced navigator who became a teacher when he left the sea. He taught astro navigation, not only to mariners but also to students of mathematics and geography and was able to demonstrate the important links between those subjects. Over many years, he developed the art of teaching navigation in an interesting way so that it could be understood by students of all ages.

For Gwenda, Jade-Marie and Russell.

Contents

Chapter 1
Introduction

In coastal waters, we can use traditional terrestrial navigation techniques to find our position in relation to geographical features with a high degree of accuracy. We do this by triangulation methods, using bearings of geographical features; sometimes combining these bearings with distances, and heights.

When out of sight of land, humans have for centuries, practiced the art of astro navigation (also known as celestial navigation) which involves similar triangulation techniques by measuring the altitudes and azimuths of celestial bodies such as stars, planets, the Sun and the Moon. (The terms azimuth and altitude are explained in chapter 2). The difficulty here of course, is that whereas in terrestrial navigation, the positions of geographical features are fixed and are precisely marked on our charts, the positions of the celestial bodies are constantly moving through space in different directions and at different speeds and cannot be marked on our charts. So how is this difficulty overcome?

Ptolemy was a Greek astronomer who lived in Alexandria in the first century AD. He believed that the Earth was at the centre of the universe as shown in the following representation of his geocentric model.

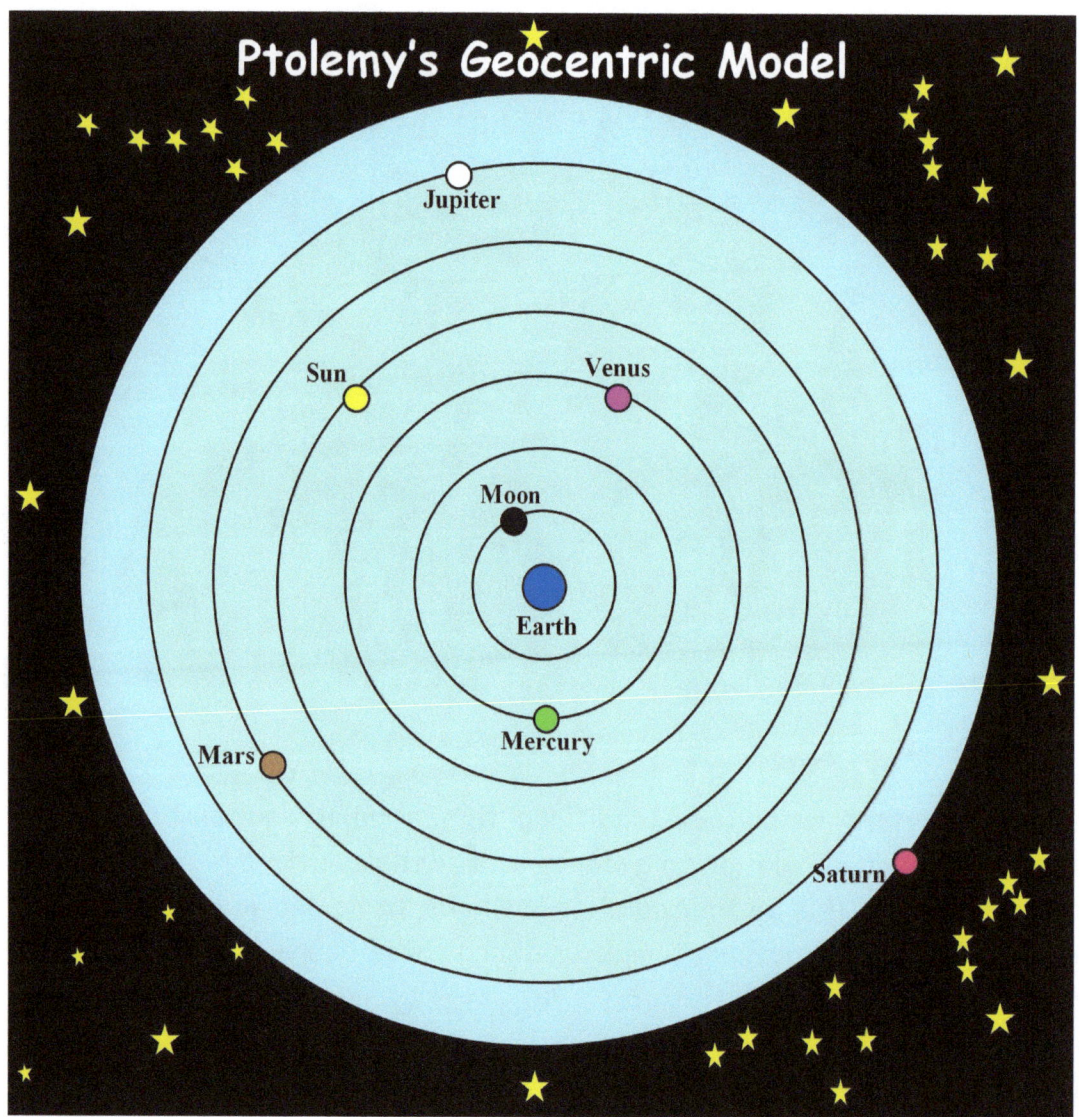

For the purposes of astro navigation, we adopt a similar Ptolemaic worldview. We assume that the Earth is at the centre of a vast sphere which we call the 'Celestial Sphere'. The 'celestial bodies' such as the Sun, Moon, stars and planets are placed on the inner surface of the celestial sphere much as we would see them in the roof of a planetarium.

The fact that the 'celestial bodies' are at greatly varying distances from the Earth and not actually on the inner surface of a sphere is not important since we are only concerned with the angular distances between them. We are not interested in the positions of celestial bodies in terms of astronomical units, we simply need to know how to locate them as they appear to us in the sky.

In the next diagram, X and Y represent two celestial bodies in space and X^1 and Y^1 represent their corresponding positions on the imaginary celestial sphere.

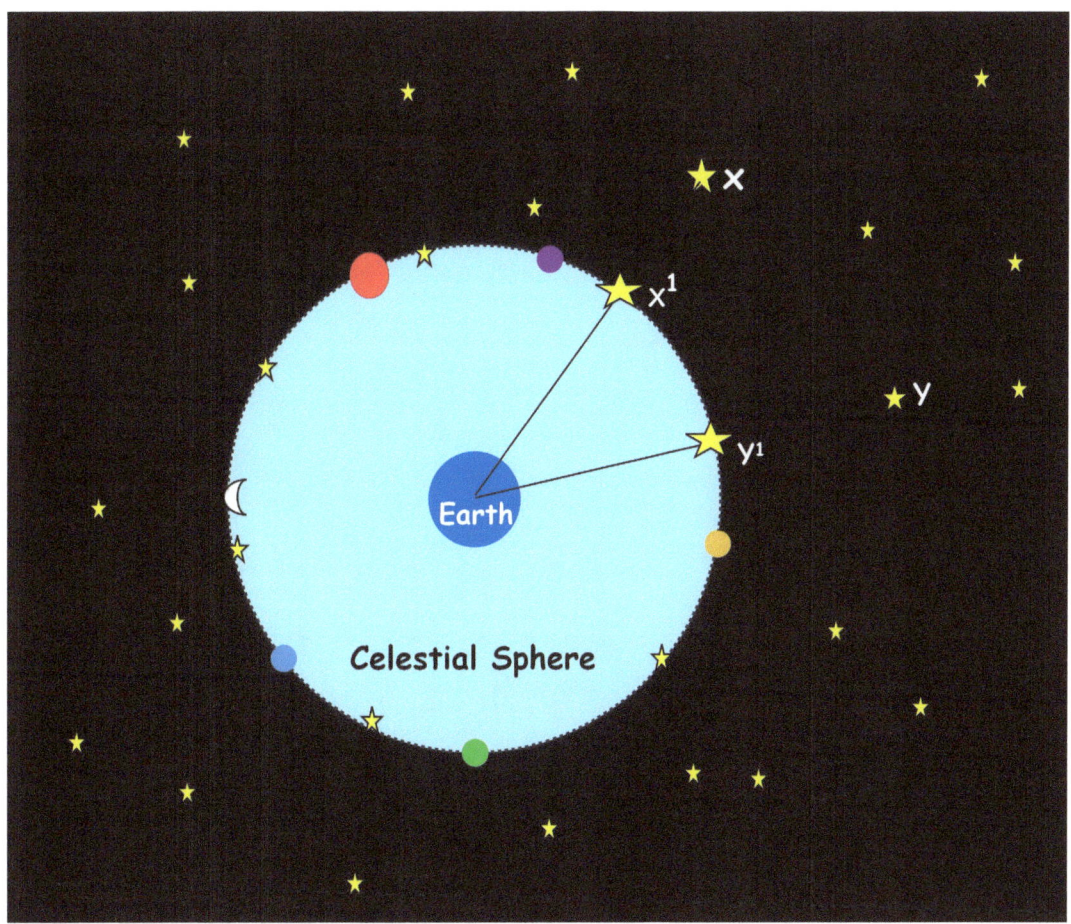

It can be seen that the angular distance between X and Y, when measured from a point on the Earth's surface, is exactly the same as that between X^1 and Y^1. So, even though the celestial bodies are really at X and Y, no error is introduced by assuming that they are at X^1 and Y^1.

Locating Celestial Bodies. A navigator will use a nautical almanac to find the Greenwich hour angle and the declination of a celestial body and use these to compute its altitude and azimuth at an assumed position through sight reduction methods. There are other methods of determining the positions of celestial bodies such as by the use of Star Globes, Star Charts etc. However, even with these, navigators must eventually locate a body by eye before its altitude and azimuth can be accurately measured at the true position.

Calculating our position on the Earth's surface by astro navigation. In the PZX Triangle diagram below, point A represents the position of an observer on the Earth's surface and point Z represents the projection of point A onto the celestial sphere. So Z is the point on the celestial sphere directly above the observer and is called the '**Zenith**

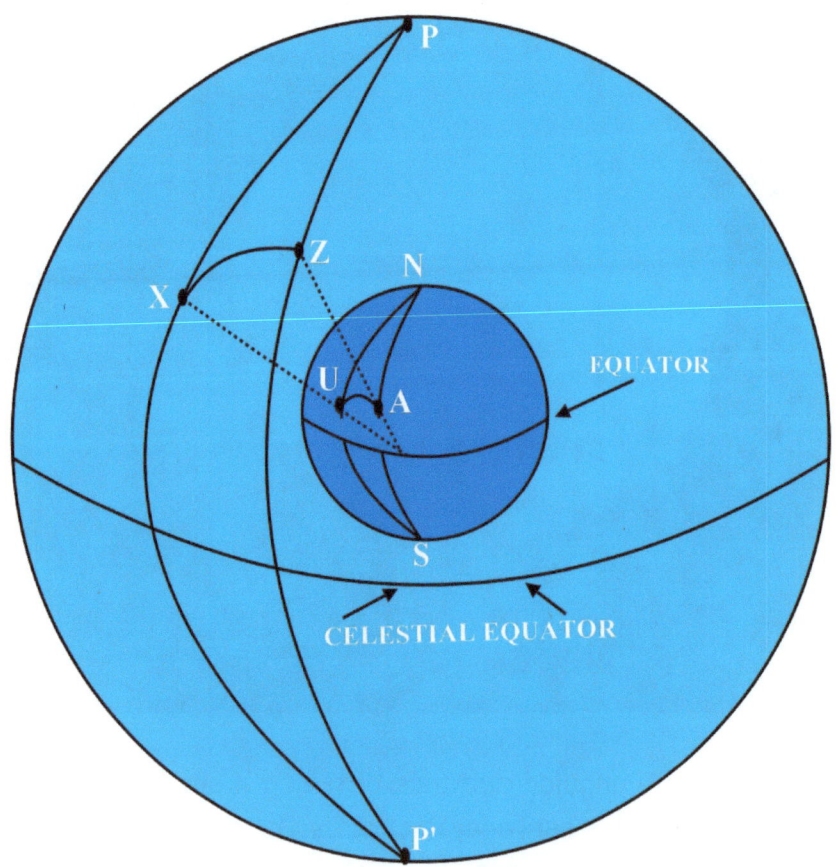

The PZX Triangle

Point X represents the position of a celestial body and point U represents its projection onto the surface of the Earth. So U is the point on the Earth's surface immediately below the celestial body and is called the 'Geographical Position' (GP).

Similarly, P is the projection of N and P' is the projection of S.

Suppose yourself to be on the Earth's surface at point A. You would not be able to see the North Pole (point N) nor would you be able to see point U (the geographical position). However, you would be able to see the celestial bodies in the sky. So, although the triangle NAU is inaccessible, you would be able to solve it, in effect, by solving the triangle PZX.

The theory of position fixing by astro navigation depends on the ability to solve the triangle PZX by relating the observed altitude and azimuth of a celestial body as measured at the true position (which is theoretically accurate) to an assumed position (which is only approximate). In this way, we are able to determine the geographical position of the celestial body and then calculate our true position in relation to it. (Note. Chapter 8 of this book gives a more detailed explanation of the methods used in Astro navigation).

The aim of this book is to look at ways in which this theory can be put to use and we begin by examining the relationships between the Earth and the other celestial bodies.

Chapter 2
The Earth and the Celestial Sphere

The Earth. The Earth is the third planet from the Sun and is the largest of the terrestrial planets which are planets with solid surfaces). (The planets are discussed in greater detail in chapter 4).

The Celestial Sphere is an imaginary sphere with the Earth located at its centre. As discussed in chapter 1, we imagine that the 'celestial bodies' such as the Sun, Moon, stars and planets are placed on the inner surface of the celestial sphere just as we would see them in the sky.

The diagram below illustrates the information about the Earth and the celestial sphere that is explained in the following paragraphs.

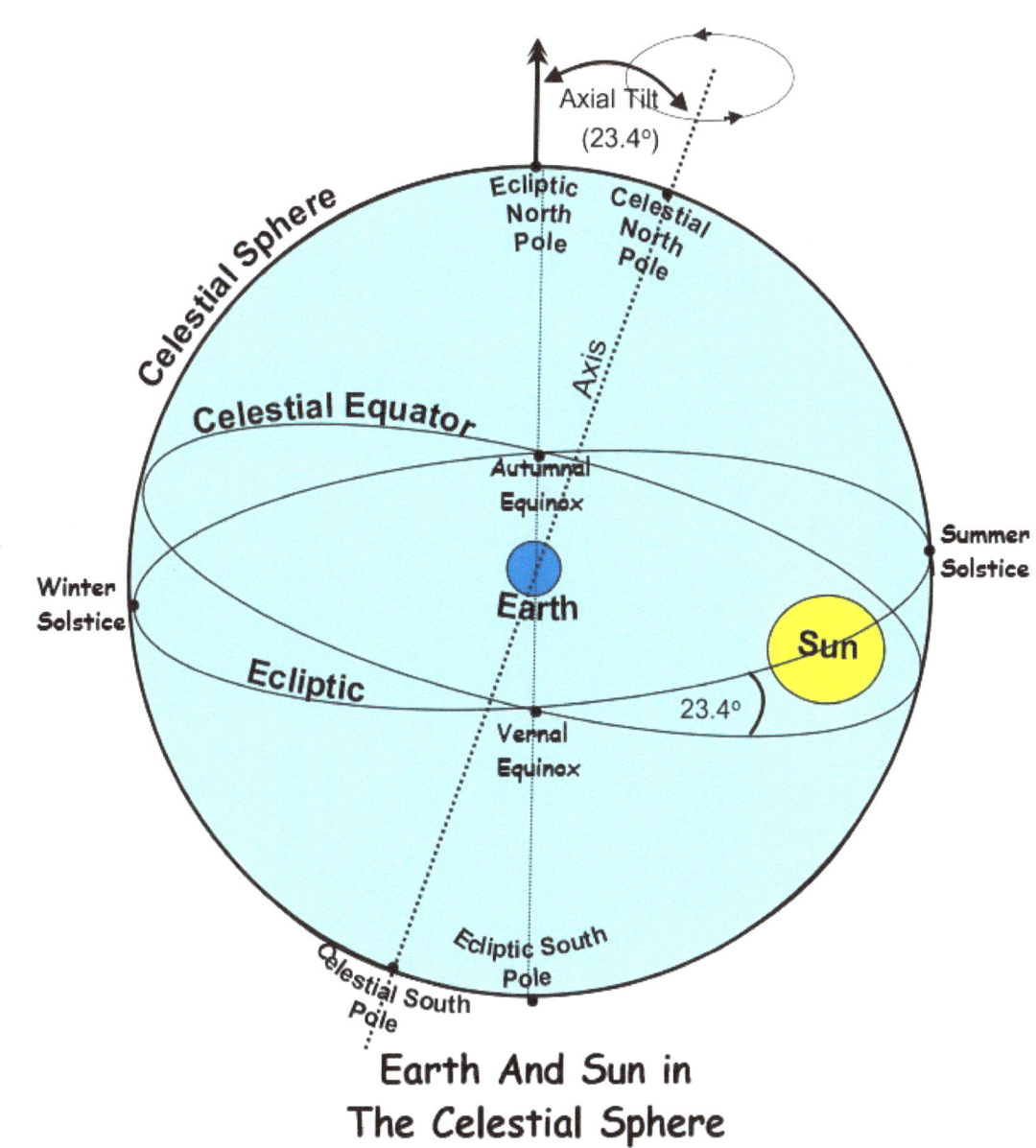

Earth And Sun in
The Celestial Sphere

Earth's Orbit.
The Earth's orbit is the elliptical path in which it travels around the Sun.
The Earth lies at a mean distance of 93 million miles (149.59787 million kilometers) from the Sun; however, the actual distance varies because of the elliptical nature of its orbit. It completes each orbit in 365.256 days (1 sidereal year). The length of the orbit is 940 million kilometers (584 million miles) and although there are slight fluctuations in the orbital speed the average is reckoned to be 30 km/s (67,000 mph).

Ecliptic. Because of the orbital motion of the Earth, the Sun appears to us to move around the celestial sphere taking one year to complete a revolution. This apparent movement of the Sun is called the Ecliptic. A year is approximately 365.25 days in length. However; for the sake of convenience, the Gregorian calendar divides three years of the cycle into 365 days and the fourth (the leap year) into 366.

Ecliptic Poles. The north and south ecliptic poles are two imaginary points where a straight line drawn from the centre of the Earth and perpendicular to the path of the ecliptic meets the celestial sphere.

Celestial Poles. The north and south celestial poles are two imaginary points where the Earth's axis of rotation meets the celestial sphere.

Geographic Poles. These are the points where the Earth's axis of rotation meets its surface. These are simply known as the North Pole and the South Pole.

Magnetic Poles. These are the north and south poles of the Earth's magnetic field and are offset slightly from the geographical poles.

Earth's Rotation. Looking down from the celestial north pole, the Earth rotates about its axis in an anti-clockwise direction or in other words, from west to east. It takes exactly 24 hours of mean time for the Earth to turn once on its axis with respect to the Sun but it takes 23 hours, 56 minutes and 4 seconds to complete one rotation with respect to the rest of the universe. The reason for the difference is that, as well as rotating on its axis, the Earth is also orbiting around the Sun and because of this, the Sun catches up with the Earth by approximately 4 minutes each day bringing the rotational period to exactly 24 hours. The amount of time it takes for the Earth to turn on its axis with respect to the universe is know as the sidereal day and the time taken with respect to the Sun is called a solar day. (Note. Mean time is explained in chapter 7).

Axial Tilt. This is the angle between the Earth's rotational axis and a line perpendicular to the ecliptic which passes through the geocentric centre of the Earth and its celestial poles. Axial tilt is measured between the celestial and the ecliptic poles. The Earth currently has an axial tilt of 23.4°; however, this is not a fixed quantity and at present, it is decreasing at a rate of about 47 arc seconds per century. This is due to a gravity induced gradual shift in the orientation of the Earth's rotational axis and is known as **Axial Precession.**

Rotational Velocity. The Earth's rotational velocity at the equator is 1,674.4 km/h. so a person standing on the equator would be travelling 1,674.4 km/h in a circle.

The Earth's equator is an imaginary line on the Earth's surface the plane of which is at right angles to the axis of rotation. It is equidistant from the North and South Poles and divides the Earth into the Northern Hemisphere and Southern Hemisphere. The length of the equator (that is the equatorial circumference) is 40,075.16 km. which is equal to 21600.5 geographical miles.

The Celestial Equator is the projection of the Earth's equator onto the surface of the celestial sphere.

Declination. The declination of a celestial body is its angular distance North or South of the Celestial Equator. The declinations of the stars change very slowly and can be considered to be almost constant for up to a month at a time.
The declination of the Sun changes relatively fast from 23.4° North to 23.4° South and back again during the course of a year.
The Moon's declination is more difficult to predict because the rate of change is even more rapid than that of the Sun and the pattern of the changes is less uniform.
Because the planets are constantly orbiting the Sun, their declinations also change rapidly in comparison with the stars.
Declination can be summarised as the celestial equivalent of Latitude and for practical reasons, we treat it as the angular distance of a celestial body North or South of the Equator.

The Equinoxes. The Sun crosses the celestial equator on two occasions during the course of a year and these occasions are known as the equinoxes. At the equinoxes, at all places on Earth, the nights and days are of equal duration (i.e. 12 hours) hence the term equinoxes (equal nights). Because the Sun is on the celestial equator at the equinoxes, its declination will of course be 0°.

The Autumnal Equinox occurs on about the 22nd September when the Sun crosses the celestial equator as it moves southwards from 23.4°N, the northernmost limit of its declination.

The Vernal Equinox occurs on about the 20th March when the Sun crosses the celestial equator as it moves northwards from 23.4°S, the southernmost limit of its declination.

The Solstices. The times when the Sun reaches the northerly and southerly limits of its path along the ecliptic are known as the solstices. The word solstice is taken from 'solstitium', the Latin for 'sun stands still'. This is because the apparent movement of the Sun seems to stop before it changes direction

The Summer Solstice (mid-summer in the northern hemisphere) occurs on about 21st June when the Sun's declination reaches 23.4° North (the tropic of Cancer).

The Winter Solstice (mid-winter in the northern hemisphere) occurs on about 21st December when the Sun's declination is 23.4°South (the tropic of Capricorn).

Note. Because each year is 365.25 days in length, the dates of the equinoxes and the solstices will vary slightly during the four-year cycle between leap years; so the Vernal Equinox sometimes falls on 20th March and sometimes on 21st. The Autumnal Equinox sometimes falls on 22nd September and sometimes on 23rd. Similarly, the Summer Solstice usually falls on 21st June but sometimes falls on 20th. The Winter Solstice usually falls on 21st December but sometimes falls on 22nd.

The tropic of Capricorn is so named because in ancient times, the Sun passed through the constellation Capricornus during the Winter Solstice on 21/22 December when the Sun's declination reached its southernmost latitude of 23.4°S. However, due to precession, the Sun is now over the constellation Sagittarius at the Winter Solstice.

The Tropic of Cancer. These days, the Sun passes through the constellation Cancer in late July; however, in the time of Ptolemy, around 2000 years ago, this occurred during the summer solstice when the Sun reached 23.4° N, the northern limit of the ecliptic. The latitude 23.4°N is still called the tropic of Cancer even though the Sun now resides in Taurus at the summer solstice.

Note. The latitude of the tropic of Cancer is currently drifting south at approximately 0.5 arc seconds per year while the latitude of the tropic of Capricorn is drifting north at the same rate.

Zenith. The **Zenith** is an imaginary point on the celestial sphere directly above the observer. It is the point where a straight line drawn from the geocentric centre of the Earth, through the observer's position and onwards, intersects with the celestial sphere.

Zenith Distance. Measuring from the Earth's centre, zenith distance is the angular distance from the observer's zenith to the celestial body's position on the celestial sphere.

Nadir. The Nadir is the direction opposite to the zenith, that is, the direction pointing directly below the observer.

Altitude and Azimuth. The diagram below illustrates the explanations of altitude and azimuth which follow.

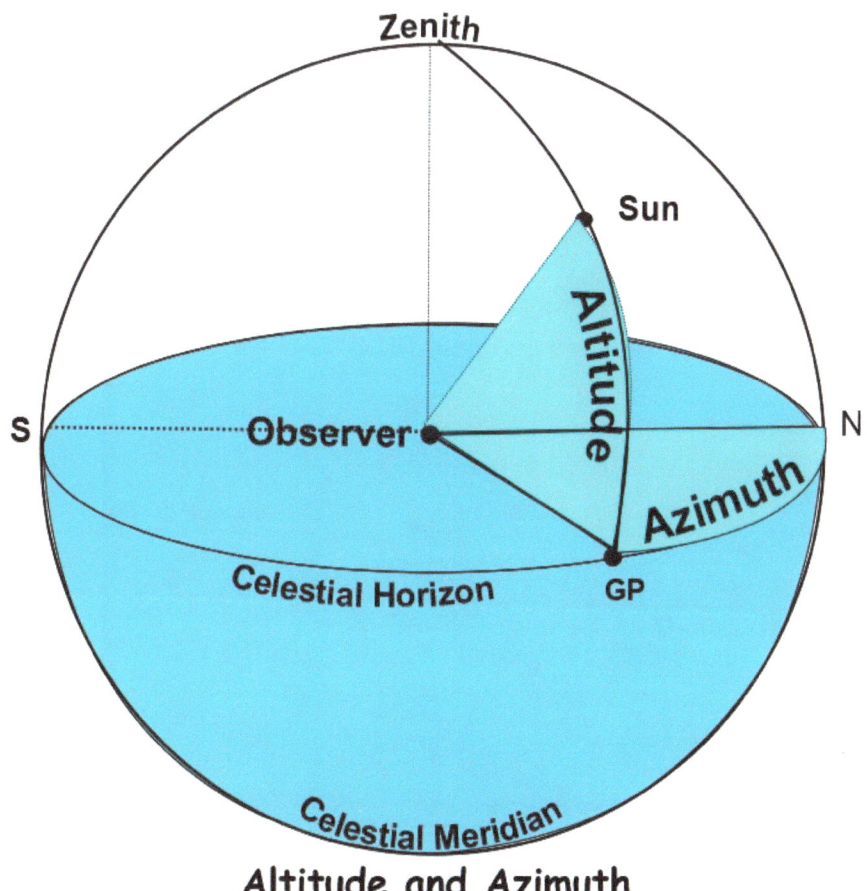

Altitude and Azimuth

Altitude. The altitude of a celestial body is the angular distance between its position in the celestial sphere and the celestial horizon as measured at the observer's position.

Azimuth. The azimuth of a celestial body is the angular distance between the observer's meridian and the direction of the geographical position of the body (GP). It is measured in the horizontal plane in relation to true north, clockwise from 0° to 360°. For example, in terms of azimuth, due east is 090° and due west is 270°. (Note. altitude and azimuth are explained in greater detail in chapter 8).

Visible Horizon (also called the Sensible Horizon). The plane of a small circle of the celestial horizon which is perpendicular to the zenith of the observer's position is called the visible horizon. In other words, it is the limit of the horizontal plane at the observer's position.

Celestial Horizon (also called the **Rational Horizon).** The celestial horizon is the plane of a great circle that passes through the Earth's centre and is parallel to the visible horizon.

Observed Altitude and True Altitude. As shown in the next diagram, the observer measures the altitude in relation to the visible horizon from his position at O on the Earth's surface. So, the **observed altitude** is the angle HOX. However, the **true altitude** is measured from the Earth's centre in relation to the celestial horizon and is the angle RCX.

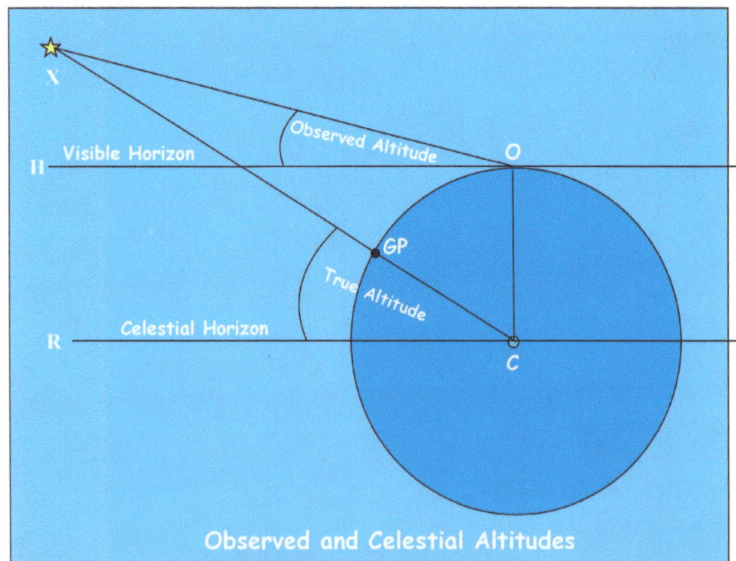

Observed and Celestial Altitudes

Local Hour Angle (LHA). In astro navigation, we need to know the position of a celestial body relative to our own position.

(*The following explanation refers to the PZX triangle diagram taken from chapter 1 and repeated below*).

LHA is the angle ANU on the Earth's surface which corresponds to the angle ZPX in the Celestial sphere. In other words, it is the angle between the meridian of the observer and the meridian of the geographical position of the celestial body (GP).

Due to the Earth's rotation, the Sun moves through 15° of longitude in 1 hour and it moves through 15 minutes of arc in 1 minute of mean time. So the angle ZPX can be measured in terms of time and for this reason, it is known as the Local Hour Angle.

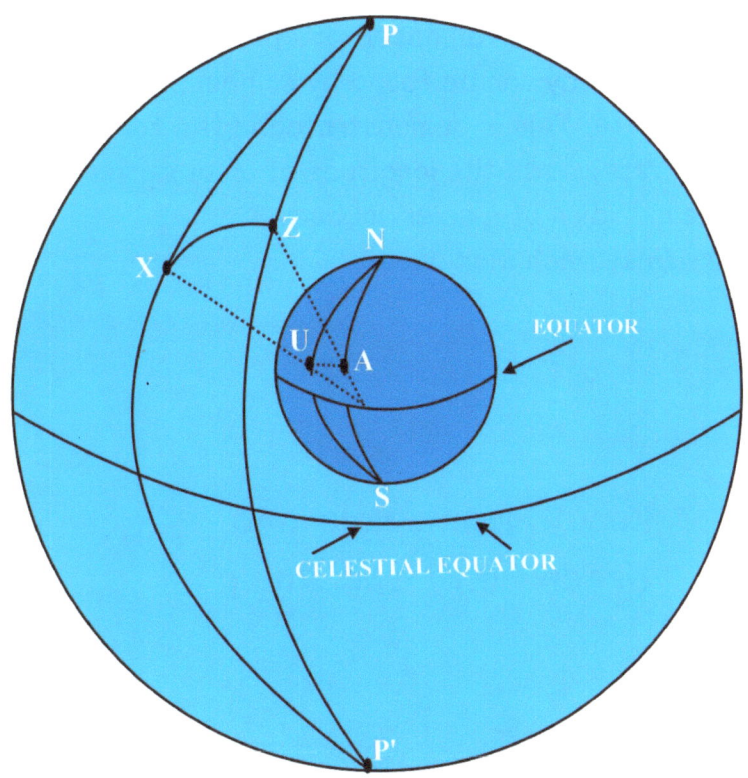

The PZX Triangle

LHA is measured westwards from the observer's meridian and can be expressed in terms of either angular distance or time. For example, at noon (GMT) the Sun's GP will be on the Greenwich Meridian (0°). If the time at an observer's position is 2 hours and 3 minutes after noon, then the angular distance between the observer's meridian of longitude and the Greenwich Meridian must be (2 x15°) + (3x 15') = 30° 45'.

Because it is after noon at the observer's position, the longitude of that position must be to the East of the Greenwich Meridian since the Earth rotates from west to east. Therefore, the observer's longitude must be 30° 45' East and since LHA is measured westwards from the observer's meridian, the LHA must also be 30° 45'. However, it should be noted that as the Earth continues to rotate eastwards, the GP of the Sun will continue to move westwards so the LHA at the observer's position will be continually changing.

Greenwich Hour Angle (GHA). As discussed above, the angle between two meridians of Longitude can be expressed as an hour angle. The hour angle between the Greenwich Meridian and the meridian of a celestial body is known as the Greenwich Hour Angle.

The Local Hour Angle between an observer's position and the geographical position of a celestial body can be found by combining the observer's longitude with the GHA. This is demonstrated in the following diagram.

In the diagram, O represents the longitude of an observer;

X represents the meridian of a celestial body;

G represents the Greenwich Meridian.

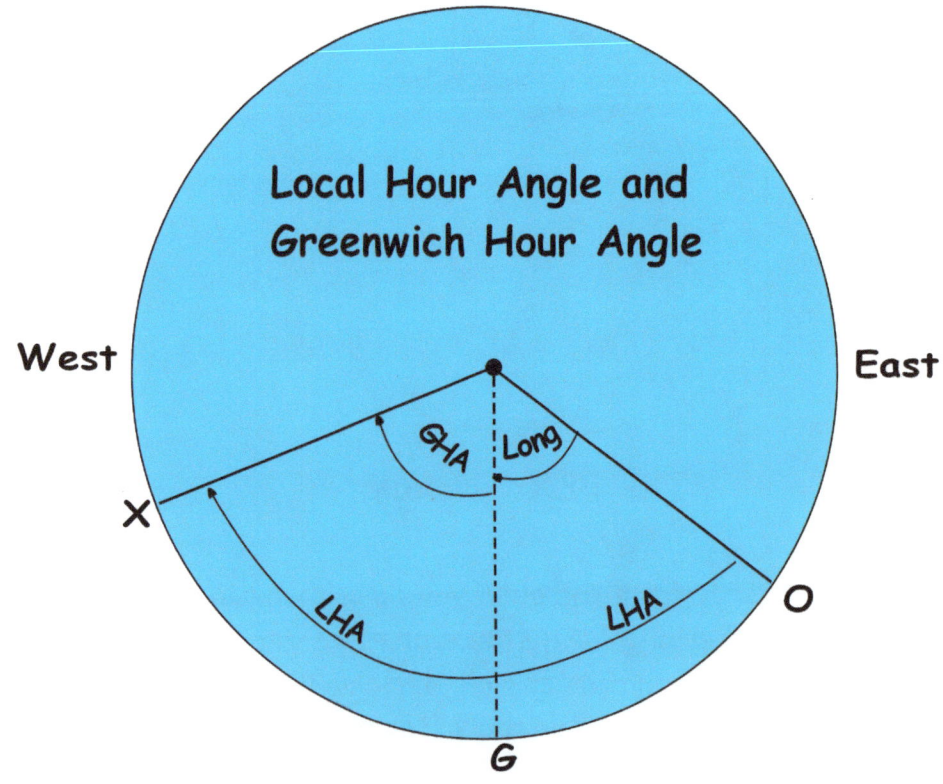

Because, in this case, the observer's longitude is east and because LHA is measured westwards from the observer's meridian to the meridian of the celestial body, LHA is equal to the GHA plus the longitude. So, if the longitude of the observer is 45°E and the GHA of the celestial body is 55°, the LHA would equal 100°.

If the longitude were to be west then this rule would change so that LHA would equal GHA minus Long.

First Point of Aries. In astronomy, we need a celestial coordinate system for fixing the positions of all celestial bodies in the celestial sphere. To this end, we express a celestial body's position in the celestial sphere in relation to its angular distances from the Celestial Equator and the celestial meridian that passes through the **'First Point of Aries'**. This is similar to the way in which we use latitude and longitude to identify a position on the Earth's surface in relation to its angular distances from the Equator and the Greenwich Meridian.

The First Point of Aries is usually represented by the 'ram's horn' symbol shown below:

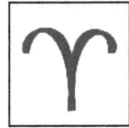

Just as the Greenwich meridian has been arbitrarily chosen as the zero point for measuring longitude on the surface of the Earth, the first point of Aries has been chosen as the zero point in the celestial sphere. It is the point at which the Sun crosses the celestial equator moving from south to north along the ecliptic (at the vernal Equinox in other words). This point is known as the 'First Point of Aries' because in 150 B.C. when Ptolemy first mapped the constellations, Aries lay in that position. However, although still named the 'first point of Aries', due to precession, the vernal equinox now lays in the constellation Pisces.

Right Ascension (RA). This is used by astronomers to define the position of a celestial body and is defined as the angle between the meridian of the First Point of Aries and the meridian of the celestial body measured in an Easterly direction from Aries. RA is not used in astro navigation; Sidereal Hour Angle is used instead:

Sidereal Hour Angle (SHA). This is similar to RA in as much that it is defined as the angle between the meridian of the First Point of Aries and the meridian of the celestial body. However, the difference is that SHA is measured westwards from Aries while RA is measured eastwards.

The following diagram illustrates the concepts discussed above.

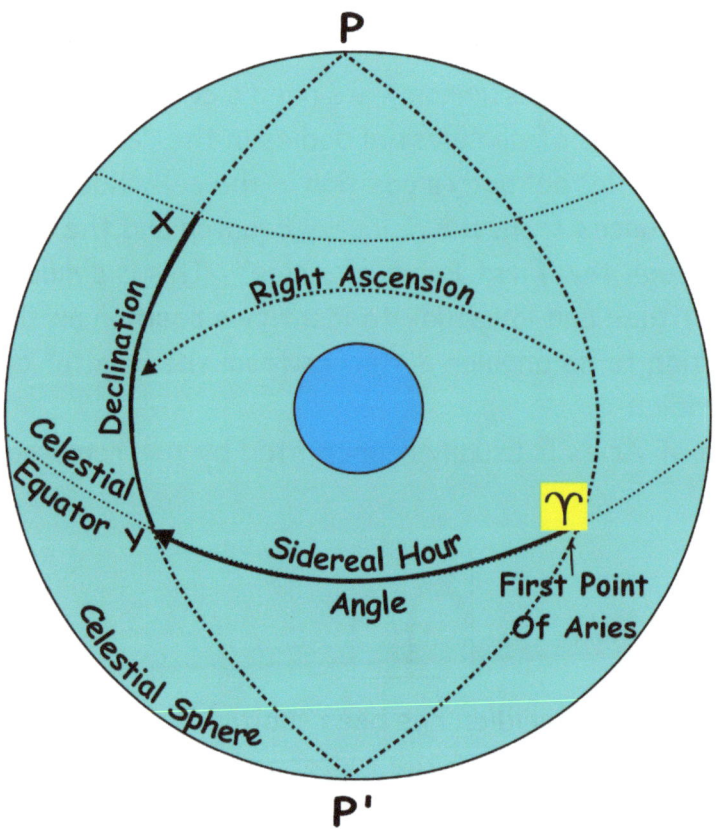

Sidereal Hour Angle, Right Ascension
and The First Point Of Aries

X is the position of a celestial body in the celestial sphere.
PXP' is the meridian of the celestial body.
Y is the point at which the body's meridian crosses the celestial equator.
♈ is the First Point of Aries.
The **Sidereal Hour Angle** is the angle ♈PY. That is the angle between the meridian running through the First Point of Aries and the meridian running through the celestial body measured at the pole P.

It can also be defined as the angular distance ⵑY. That is the angular distance measured **westwards** along the Celestial Equator from the meridian of the First Point of Aries to the meridian of the celestial body.

Right Ascension can also be defined as the angle between the meridian of the First Point of Aries and the meridian of the celestial body but the difference is that it is measured in an **easterly** direction from Aries.

From this, we can conclude that

RA   = 360° – SHA and

SHA = 360° - RA.

The Importance of Altitude, Azimuth and Zenith Distance in Astro Navigation. Please consider the following diagram. The celestial sphere is drawn in the plane of the observer's meridian with the observer's zenith (Z) at the top.

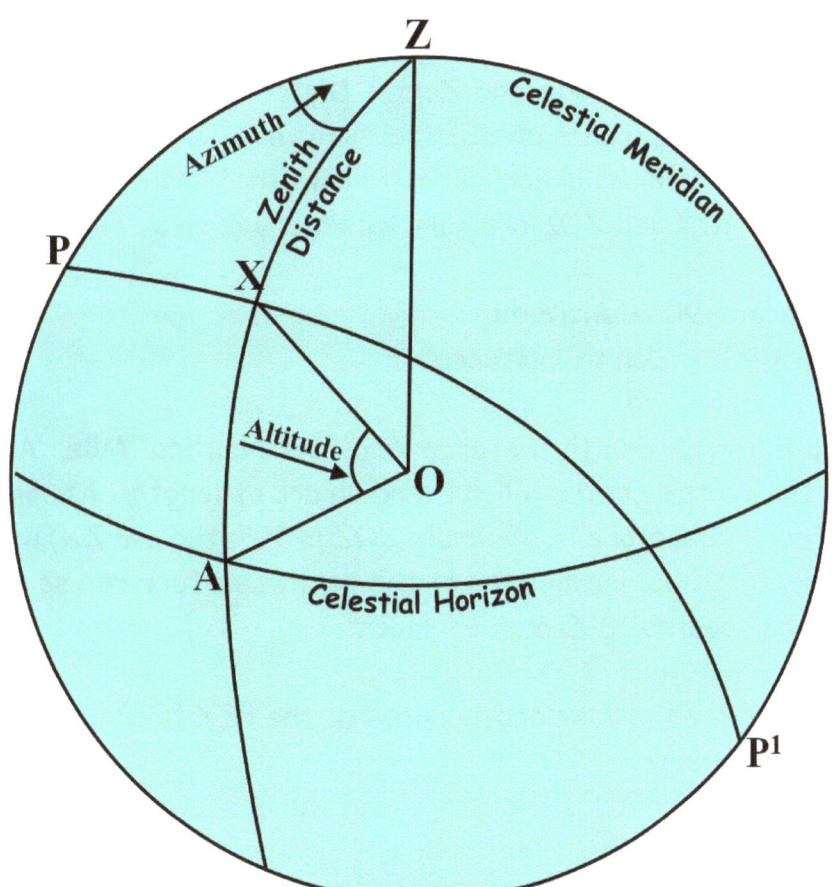

Altitude, Azimuth and Zenith Distance

Point O represents both the observer and the Earth.

Z represents the observer's zenith.

X is the position of a celestial body in the celestial sphere.

A is the point where the virtual circle running through the position of the celestial body meets the celestial horizon.

P and P^1 are the north and south poles respectively.

The Zenith Distance. In this diagram, the zenith distance is the angular distance ZX that is subtended by the angle XOZ and is measured along the vertical circle that passes through the celestial body. (A vertical circle is a great circle that passes through the observer's zenith and is perpendicular to thé celestial horizon).

The Altitude. Altitude is the angle AOX, that is the angle from the celestial horizon to the celestial body and is measured along the same vertical circle as the zenith distance.

Relationship between Altitude and Zenith Distance

Since the celestial meridian is another vertical circle and is therefore, also perpendicular to the celestial horizon, it follows that angle AOZ is a right angle and angles AOX and XOZ are complementary angles. From this we can deduce that:

Zenith Distance = 90° – Altitude

and **Altitude = 90° – Zenith Distance**

Relationship Between Zenith Distance And The Nautical Mile. An angle of 1 minute at the earth's centre will subtend an arc of length 1 n.m on the earth's surface. Therefore if the angle XOZ is 30° (the arc ZX) will be equal to 30 x 60 = 1800 arc minutes at the earth's surface and so the zenith distance will be equal to 1800 nautical miles.

At this point, we need to take another look at the PZX triangle diagram which is repeated below.

In this diagram the arc AU is the arc joining the observer's position to the geographical position of the celestial body. This arc when projected onto the celestial sphere forms the arc ZX which is the zenith distance. Therefore, from the discussion above, it can be seen that the angular distance ZX is equal to the angular distance AU which when converted to nautical miles will give us the distance from the GP of the body to the position of the observer.

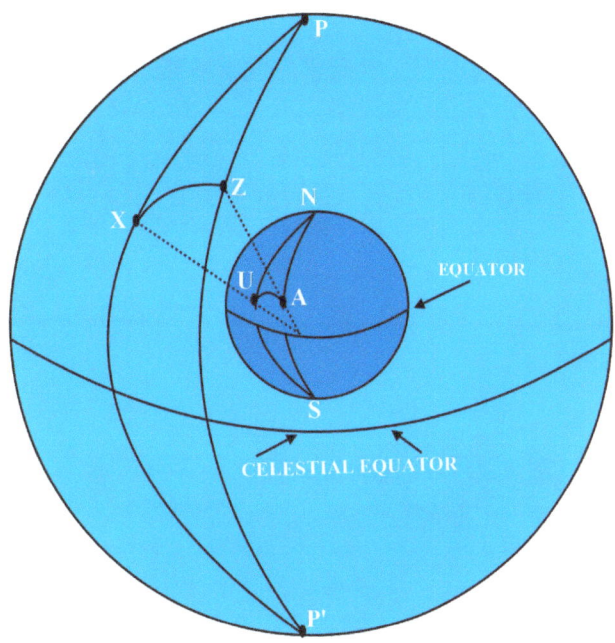

The PZX Triangle

Azimuth. The angle PZX is the azimuth of the celestial body and is the angular distance between the observer's celestial meridian and the direction of the geographical position of the body.

Summarizing The Role Of Altitude, Azimuth And Zenith Distance In Astro Navigation. The preceding discussion illustrates the importance of altitude and azimuth in astro navigation. It can be seen that by measuring the altitude of a celestial body, we are able to easily calculate the zenith distance which will give us the distance in nautical miles from the observer's position to the geographical position of the body. The azimuth will give us the direction of the geographical position of the celestial body from the observer's position. This explains why measuring the altitude and azimuth are the first steps in determining our position in astro navigation. (A thorough treatment of this topic can be found in the book Astro Navigation Demystified).

Chapter 3
The Moon

The Moon, our only natural satellite, is 27% the size of the Earth and has a circumference of 10,917 km. Its average distance from the Earth is 384,400 Km. but it is drifting away at approximately 3.7 cm. per year. Its orbital length is approximately 2,412,517.5 km. and it completes each orbit in 27.3 Earth days.

The Dark Side of the Moon. Because the Moon rotates around its own axis at the same time that it takes to orbit the Earth, the same side always faces the Earth and is lit by sunlight. For that reason, we can only ever see one side of the Moon from the surface of the Earth. The other side of the Moon is in perpetual darkness and was not seen by humans until 1968 when Jim Lovell, Frank Borman and William Anders went into orbit around it in the Apollo 8.

PHASES OF THE MOON

The diagram below shows, that as the Moon completes its 27.3 day orbit around the Earth, we see it pass through various phases of illumination. It goes from New Moon, to Full Moon and back to new Moon again.

MOON PHASES

The Phases.

New Moon. When the illuminated side of the Moon is facing away from the Earth, the Moon and the Sun are lined up on the same side of the Earth, so we can only see the shadowed side.

During a new moon, we can see no reflected sunlight from the Moon but we can see sunlight which has been reflected from the Earth and re-reflected in the Moon; this is known as earthshine.

Waxing Crescent – The waxing crescent moon begins with the first sliver of a crescent that we can see after the new moon and gradually increases in width until it reaches the first quarter. From the northern hemisphere, the crescent moon has the illuminated edge of the Moon on the right. This situation is reversed for the southern hemisphere. "Waxing" means that the Moon becomes more illuminated night-by-night,

First Quarter – This occurs when the Sun and the Moon make a 90-degree angle compared to the Earth. Although it's called a quarter moon, we actually see it as half illuminated.

Waxing Gibbous – This phase of the Moon occurs when more than half of the Moon is illuminated but it is not yet a full Moon.

Full Moon – This is the phase when the Moon is brightest in the sky. The Moon and the Sun are lined up on opposite sides of the Earth, so from our perspective here on Earth, the Moon is fully illuminated by the light of the Sun.

Waning Gibbous – This phase occurs between the full moon and the last quarter when more than half of the Moon is illuminated but it is gradually decreasing in width. (The term "waning" means that it is getting less illuminated each night).

Last Quarter – At this point of the lunar cycle, the Moon has reached half illumination again. Now it is the left-hand side of the Moon that's illuminated, and the right-hand side is in darkness (from a northern hemisphere perspective).

Waning Crescent – Between the last quarter and the new moon, the Moon becomes a crescent again and continues to decrease in width until it becomes a final sliver of light again before going into full darkness.

The Ocean Tides.

The rise and fall of the ocean tides is caused by the gravitational forces of the Sun and the Moon.

Tidal Effects of the Moon. If it were not for the gravitational attraction of the Sun and the Moon, the water level of the seas and oceans would be kept at equal levels by a combination of the Earth's own gravity pulling it inwards and centrifugal force pushing it outwards. However, the gravitational force of the Moon is strong enough to attract the water towards it and cause a bulge beneath it. As the Earth rotates and the Moon orbits around it, the bulge follows the Moon causing high tides in its vicinity. The combined effects of the Earth's rotation and the Moon's orbit around it cause the 'bulge' to move around the Earth in 24 hours and 50 minutes and so it would seem, at first sight, that we would get high tides only at that time interval. However, other forces are at play. On the opposite side of the Earth to where the 'bulge' occurs, the Moon's gravitational pull is at its weakest and this allows the Earth's centrifugal force to push the water outwards and so cause another bulge therefore giving us two high tides a day. This means that the time between high tides is approximately 12 hours and 25 minutes and the time between high tide and low tide is 6 hours 12.5 minutes in deep ocean areas. This can change dramatically owing to a variety of factors such as the topography of the ocean floor, local currents, varying water depths and the declination of the Moon.

The height of the tides vary because the Moon is not always at the same distance from the Earth due to its elliptical orbit. As the Moon comes closer to the planet, its gravitational pull increases and this leads to higher tide levels. Conversely, when the Moon's orbit takes it further away from the Earth, the tides become lower. When the Moon is at its closest distance to the Earth, its gravitational pull increases by as much as 50% and this leads to higher sea levels on Earth. When it is at its furthest distance, sea levels are much lower,

Tidal Effects of the Sun. The Sun also affects the rise and fall of the tides on Earth. The gravitational attraction of the Sun pulls the ocean water towards it but at the same time, the effect of the Earth's rotation around the Sun creates a centrifugal force which pushes the water outwards on the side facing away from the Sun. The combined effect of these two forces creates a tidal bulge on the side of the Earth facing away from the Sun. However, this effect of the Sun is less than that created by the Moon which is much closer to the Earth.

Spring Tides. When the Sun, the Moon and the Earth are in syzygy, that is when they are lined up as during a Full Moon or New Moon, the combined tidal effect of the Sun and Moon is at its greatest and causes what is known as Spring Tides. This has nothing to do with the season of Spring but to do with the saying that the water 'springs' higher than normal.

Neap Tides. When the directions of the Sun and the Moon in relation to the Earth are at right angles, as during a Quarter Moon, the combined effects of their gravitational pull is less and so the height of the tides is much lower and are known as Neap Tides.

Spring Tides

(Sun and Moon Aligned - Highest High Tides)

Bulge caused by gravitational pull of Sun

Bulge caused by gravitational pull of Moon

Bulge caused by combined gravitational pull of Sun and Moon

Bulge caused by Centrifugal Force of Earth

Neap Tides

(Sun and Moon Alignment at Right Angles With Respect To The Earth - Lowest High Tides)

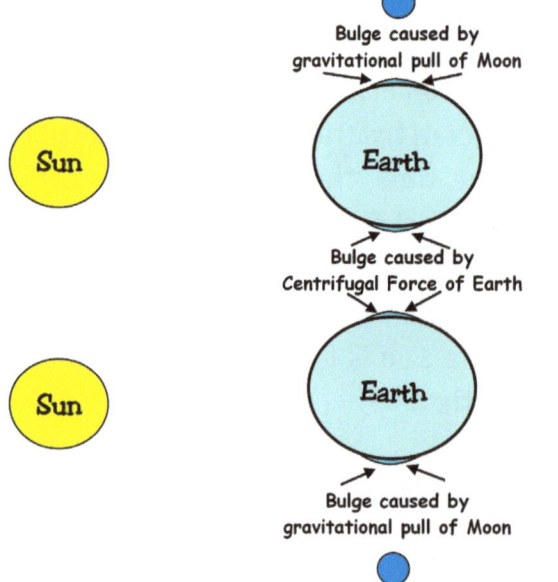

Bulge caused by gravitational pull of Moon

Bulge caused by Centrifugal Force of Earth

Bulge caused by gravitational pull of Moon

Measuring the Altitude of the Moon.   When a navigator measures the altitude of the Moon, there are several corrections that he has to make to the readings.

Corrections For The Moon's Semi-Diameter. The point on the Moon's circumference nearest to the horizon is called the **lower limb** and the point furthest from the horizon is called the **upper limb**.

When the Moon is not full, sometimes only the upper limb will be visible and sometimes only the lower limb.

From the diagram below it can be seen that sometimes, depending on the phase of the Moon, either the upper or the lower limb cannot be seen.

Waxing Crescent Moon   Waxing Gibbous Moon

Waning Crescent Moon   Waning Gibbous Moon

It should be noted that whether the Moon's upper or lower limb is visible is dependent not only on its phase but also on the relative altitudes of the Sun and the Moon. For example, if, one morning, a crescent or gibbous moon is visible in the eastern sky and the Sun is at a higher altitude, only the upper limb will be visible but if, in the evening of the same day, the Moon is visible in the western sky and the Sun has set below the western horizon, only the lower limb will be visible.

In navigational practice, the altitude that we measure is that of the lower limb; however, when the lower limb cannot be seen, we have no choice other than to measure the altitude of the upper limb.

Regardless of which limb we use, what we really need is the altitude of the Moon's centre so we must either add or subtract the value of its semi-diameter.
The following diagram shows why the semi-diameter must be added when the altitude of the lower limb is measured.

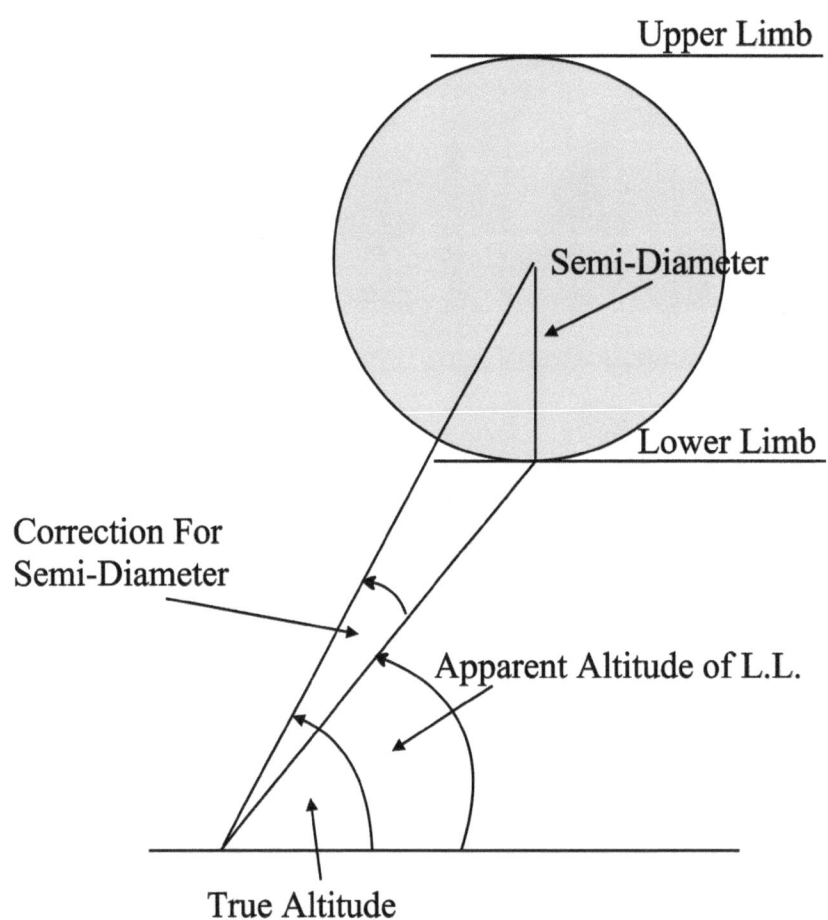

Semi-Diameter From Lower Limb

As the Moon travels around its orbit and its distance from the Earth changes, so the value of the visible moon's semi-diameter will change. The value of the Moon's semi-diameter for each day is given in the daily pages of the Nautical Almanac.

Corrections to Sextant Readings. A number of corrections have to be made to the sextant altitude before we arrive at the **True Altitude.**

Corrections For Refraction. When a ray of light from a celestial body passes through the Earth's atmosphere, it becomes bent through refraction and this causes the apparent (observed) altitude to be greater than the true altitude. Since the sextant measures the apparent altitude, a correction for refraction must be applied to find the true altitude. Refraction is at its greatest when the altitude is small (i.e. when the celestial body is near the horizon) and becomes less as the altitude increases.
The effects of refraction are illustrated in the diagram below.

Effect of Refraction on Altitude

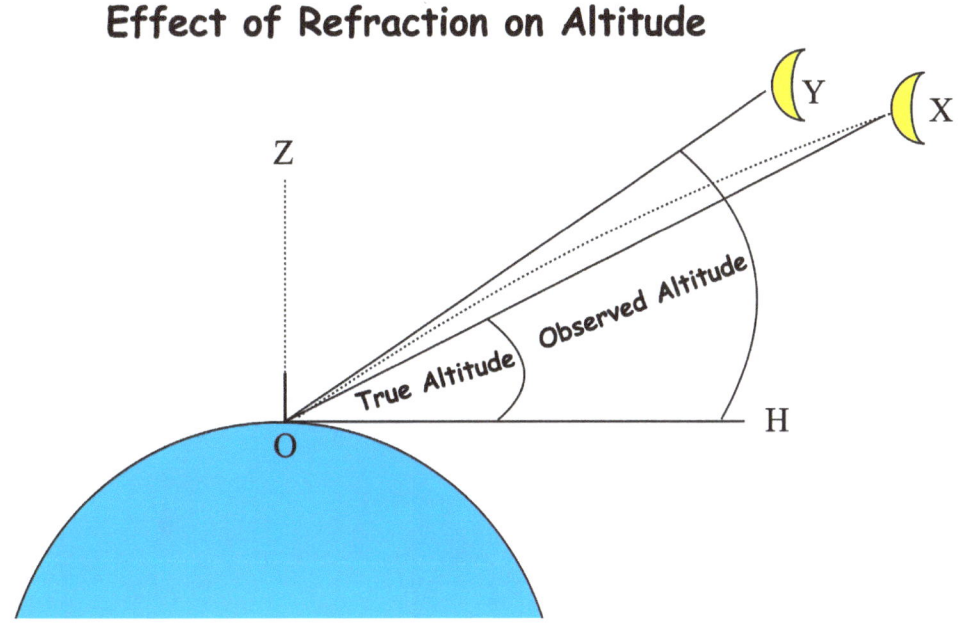

O is the observer's position and Z is the zenith at that point.
OH is the horizon
XOH is the true altitude of the Moon from the observer's position.
However, due to refraction, the celestial body appears to be at Y and so

YOH becomes the observed altitude and a correction will have to be made to compensate for this.

Corrections For Parallax. We measure the altitude of a celestial body from our position in relation to our visible horizon; this is known as the **observed altitude**. However, when calculating the **true altitude**, measurements are made from the Earth's centre in relation to the celestial horizon. The displacement between the observed position of an object and the true position is known as **parallax**.

Parallax corrections for the Moon. Because the Sun and the Moon are relatively close to the Earth, parallax will be significant and so a correction has to be made. These corrections are included in the altitude correction tables in the Nautical Almanac.

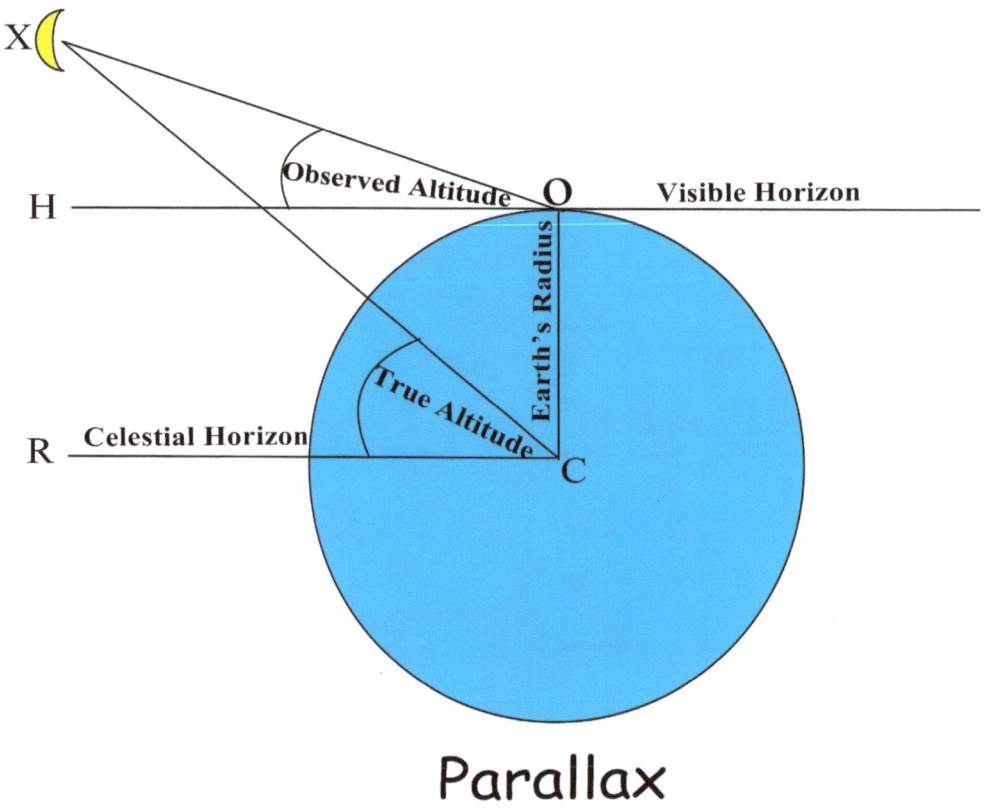

Parallax

Horizontal Parallax. Parallax error is greatest when the celestial body is close to the horizon and decreases to zero as the altitude approaches 90°. It is negligible except in the case of the Moon which is close to the Earth in comparison with the other celestial bodies. Because horizontal parallax is significant in the case of the Moon, a separate correction has to be applied. The hourly values of horizontal parallax for the Moon are listed in the daily pages of the Nautical Almanac.

Dip. A correction has to be made to allow for the height of the observer's eye above the horizon; this is known as Dip.

Consider the diagram below:

O is an observer's position on the Earth's surface and E is the position of his eye. We can see that, as the observer's height of eye is raised above sea level, his visible horizon 'dips' below the true horizon and so the altitude measured at E becomes greater than that measured at O.

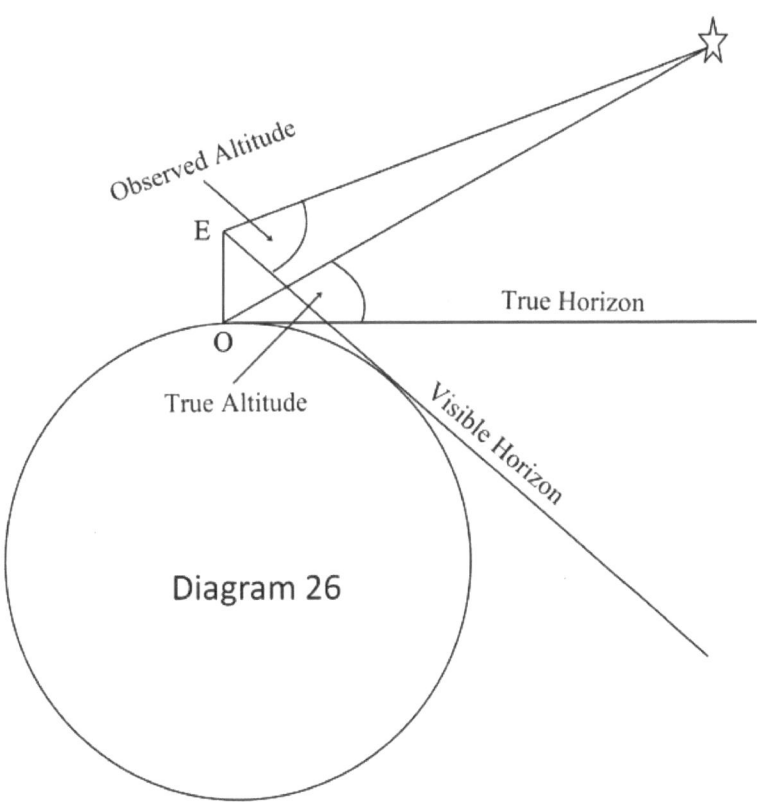

Diagram 26

Dip is the error caused by this difference and has to be subtracted from the sextant reading.

Chapter 4
The Planets

There are 8 planets in our solar system and in order of their distances from the Sun, they are Mercury, Venus, Earth, Mars, Jupiter, Saturn, Uranus and Neptune.

The picture above includes Pluto which was classified as a planet until 2006; however, it is no longer considered to be one, even though it orbits the Sun in the same way that the planets do. Instead it is now classified as a dwarf planet and is relatively small compared to the existing planets, It is primarily made of rock and ice, and is only about 1/6 the mass of our Moon.

Looking from the surface of the Earth, five of the other planets: Mercury, Venus, Mars, Jupiter and Saturn can be seen by the naked eye but Neptune and Uranus can only be seen with the aid of a telescope or binoculars.

The Inner Planets.

Mercury, Venus, Earth and Mars orbit relatively close to the Sun and are therefore known as the Inner Planets. They are also known as the **Terrestrial Planets** because they consist mostly of rocks and metals and have solid surfaces. They all have atmospheres which range from very thick on Venus to very thin on Mercury.

The Outer Planets.

Jupiter, Saturn, Uranus and Neptune are known as the outer planets. They are, on average, ten times the mass of the Earth and are composed mostly of gases such as hydrogen and for these reasons, they are also known as the Gas Giants.

The Dwarf Planets.

A dwarf planet is a celestial body that orbits the Sun but does not have sufficient mass to clear other objects out of its path in the way that the planets do.

There are 5 officially recognized dwarf planets in our solar system and these are Ceres, Pluto, Haumea, Makemake, and Eris. Ceres is located in the asteroid belt and is the largest of the dwarfs with Pluto, which is found in the outer solar system, taking second place.

The Navigational Planets. Of the seven planets in our solar system excluding Earth, only those that are sufficiently prominent to be observed with an ordinary sextant are considered to be 'navigational planets' and these are **Venus, Mars, Jupiter and Saturn.** Because this book is concerned with the application of astronomy to navigating on the surface of the Earth, we will focus our attention on just the navigational planets and their relationships with the Earth for the rest of this chapter. (Planet Earth is discussed in great detail in chapter 2).

Venus.

Venus is the second largest terrestrial planet, the Earth being the largest. It is named after the Roman goddess of love and beauty, and is sometimes referred to as the Earth's sister planet due their similar size and mass. Venus is the closest planet to the Earth; however, the distance between the two planets can vary by as much as 223 million km due to the fact that their elliptical orbits are not concentric. The closest that Venus can approach the Earth is 38 million km and at its farthest, it can be as far away as 261 million km.

After the Sun and the Moon, Venus is the third brightest celestial body in our night sky and for that reason, it is an important navigational planet.

Venus – Evening Star or Morning Star?

Venus sometimes appears as an 'evening star' above the western horizon shortly after sunset and sometimes appears as a 'morning star' above the eastern horizon shortly before sunrise. In primitive times, people regarded the 'evening and morning stars' as two different heavenly bodies but in the sixth century BC, Pythagoras suggested that they might be one and the same body.

Ptolemy believed that the Earth was the centre of the Universe and that the Sun moved in a circular orbit around the Earth which was stationary. He also believed that Venus orbited the Sun as the Sun itself orbited the Earth and that this explained why Venus appeared from Earth as an 'evening star' during part of its orbit and as a 'morning star' during another part. The Ptolemaic hypothesis can be explained with the aid the diagram below:

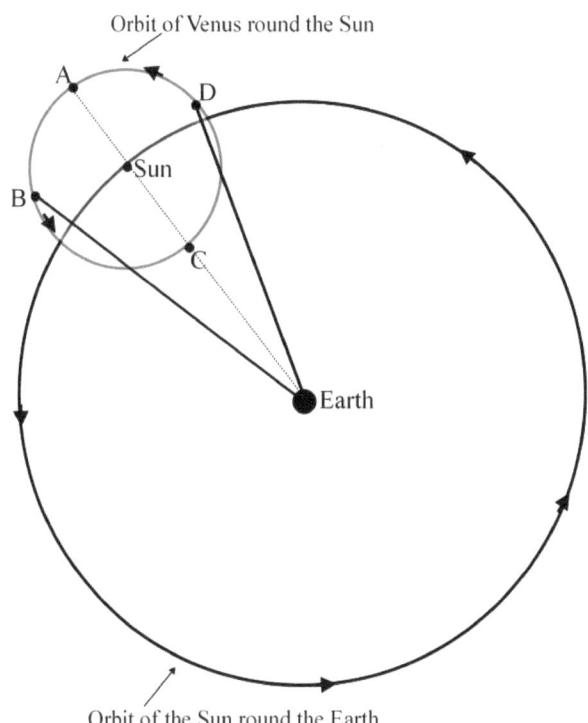

According to the Ptolemy, if Venus were at a point in its orbit somewhere on the semicircle ABC, say at B, then the line joining the Earth to Venus would be to the left of the line joining the Earth to the Sun. Therefore, looking from Earth, Venus would appear to be to the left of the Sun so, when the Sun set in the west, Venus would be seen for some time after sunset above the western horizon as an 'evening star'. In a similar way, if Venus were at a point somewhere on the semicircle ADC, say at D, it would appear from Earth to be to the right of the Sun and it would therefore appear above the eastern horizon just before sunrise as a 'morning star'.

We now know, thanks to the work of Copernicus, that both Earth and Venus orbit the Sun. According to the Copernican hypothesis, the reason that

Venus sometimes appears as an 'evening star' and sometimes as a 'morning star' can be explained with the aid of another diagram:

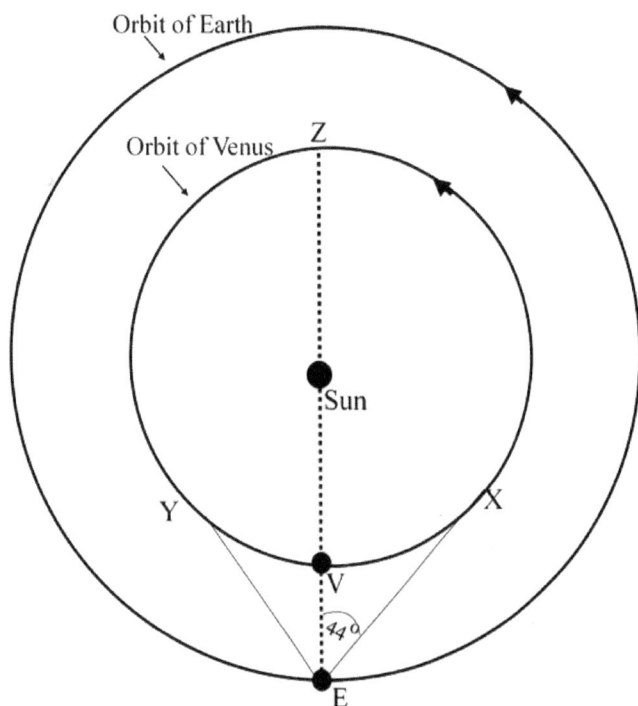

The average distance of Venus from the Sun is 108 million Km. while the average distance of Earth from the Sun is 150 million Km. Because the circumference of the orbit of Venus is smaller than that of the Earth, we can safely assume that Venus takes less time to complete its orbit than the Earth does. Whereas the Earth moves through 360° in 365.25 days, Venus completes its orbit in 225 days. For this reason Venus gains 0.6° per day on the Earth and overtakes it at intervals of approximately 600 days. From the diagram it will be seen that, after Venus has overtaken the Earth (that is after it has passed the point V) it becomes a 'morning star'. It increases its angular distance to the right of the Sun until it reaches 44° at point X. After passing X, the angular distance to the right of the Sun decreases until it reaches point Z, which is behind the Sun. After passing Z, it makes its appearance to the left of the Sun and is now an 'evening star'. As it approaches Y, its angular distance left of the Sun increases until it reaches 44° at point Y.
After passing Y on its way to V, its angular distance left of the Sun decreases until, at V, it is zero when it will pass either slightly above or

below the Sun. On extremely rare occasions, it crosses in front of the Sun and this is known as a 'transit of the planet'.

To sum up, Venus overtakes the Earth at intervals of approximately 600 days. During approximately 300 of these days, it is a 'morning star' and for the other 300 days, it is an 'evening star'. The maximum angular distance right or left of the Sun, is roughly 44°.

What about Mercury? In a similar way, Mercury is also an 'evening star' and a 'morning star'. Mercury is the closest planet to the Sun and because of its elliptical orbit, its distance from the Sun ranges from 29 million Km. to 47 million Km. Because the circumference of its orbit is comparatively small, it gains 3° on the Earth per day and overtakes it on average every 120 days. For 60 of these days it will be a 'morning star' and for the other 60 it will be an 'evening star'. Its maximum angular distance left or right of the Sun is roughly 24°.

So Venus and Mercury are both 'morning stars' and 'evening stars' but it is quite easy to tell them apart for the following reasons: Venus usually appears much higher in the sky than Mercury and is far brighter.

Mars, 'The Red Planet'.

Mars is named after the Roman god of war. It is the fourth planet from the Sun and is one of the terrestrial planets. Its surface is composed mostly of iron oxide which gives it a red colour and this makes it very easy to locate in the sky so adding to its value as a navigational planet.

The mean distance of Mars from the Sun is 227.9 km (141.71 million miles) which is 78.31 million km greater than the mean distance from the Earth to the Sun. It is about half the size of Earth in diameter and it rotates about its axis in 24 Earth hours, 37 minutes, 23 seconds.

Like the other planets, Mars orbits the Sun in an eccentric, elliptical orbit so the distance to Mars from Earth is constantly changing. The closest the planets could come together is 54.6 million km (33.9 million miles) and the farthest apart they can be is about 401 million km (250 million miles).

The orbital path of Mars is more complex than those of Mercury and Venus and we can best understand it by considering its motion relevant to the distant stars. To do this, we must study the effect of the motions of both the Earth and Mars around the Sun.

Firstly, the circumference of Mar's orbit is 1.6 times greater than that of the Earth and with an orbital speed of 24 km/s, it takes 686.98 Earth days to complete each orbit while Earth travels at 30 km/s taking 365.25 days. Thus, the Mars' year is 1.8 times as long as the Earth year.

Secondly, the orbit of the Earth is closer to the Sun than that of Mars, and this puts Earth on the inside track, so to speak. Taking this into account along with Earth's greater speed, it is easy to see that the Earth will overtake Mars from time to time; in fact it laps Mars every 26 months.

The next diagram represents the orbits of the Earth and Mars around the Sun with the Earth on the 'inside track' and Mars on the outside.

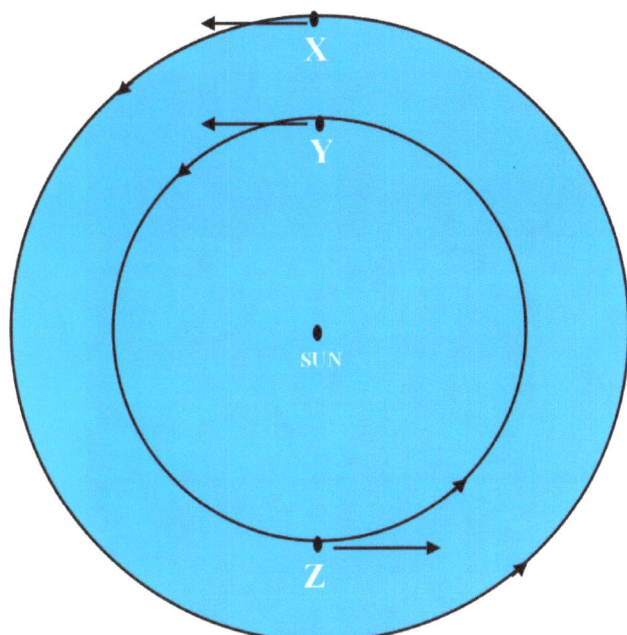

Mars Retrograde Motion

When Mars is at point X and Earth is at point Y, Mars and Earth are at their closest and although they are both moving from right to left of the diagram, because Mars is travelling 6 km/s slower, it will appear, to an observer on the Earth, to be moving left to right; this is **retrograde motion.** However, if the Earth is at point Z, it will appear that Mars is moving in the opposite direction, that is right to left. The speed at which the two planets will be moving in opposite directions will be equal to their combined orbital speeds; i.e. 54 km/s. This is called **prograde motion'.**

These changes in the apparent motion of Mars from retrograde to prograde and vice versa are not sudden changes. Before a change in direction, the planet seems to slow down and then pause for about a week before starting to move in the new direction.

To summarize, Mars will sometimes appear to be moving from left to right with respect to the background stars, sometimes it will seem to move in the opposite direction and in between these changes in, it will appear to pause.

Jupiter

Aptly named after the Roman 'King of the Gods', Jupiter has 317 times the mass of the Earth and two and a half times the mass of all the other planets in our solar system combined. It is the fifth planet out from the Sun and is one of the outer planets. It is also one of the 'gas giants' being largely composed of gases, primarily hydrogen. The distance of Jupiter from the Sun is constantly changing because, like the other planets, its orbit is elliptical; the mean distance is 778,300,000 km. It travels a distance of 4,887,600,000 km at a speed of 13 km/s to complete each orbit taking 4,332.82 Earth days or 11.85 Earth years to do so.

Jupiter appears sometimes an 'evening star' and sometimes as a 'morning star' but not for the same reasons as Venus. Jupiter will disappear behind the Sun when it is in conjunction with the Earth, that is when the two planets are on opposite sides of the Sun. As it approaches conjunction, Jupiter appears as an 'evening star' near the western horizon at twilight. When it emerges from behind the Sun, it appears as a 'morning star' near the eastern horizon. This process is repeated at intervals of approximately 13 months when Jupiter and Earth approach conjunction.

Just as the movement of Mars appears to change between retrograde and prograde, so does that of Jupiter. The topic of retrograde motion was dealt with in our discussion of Mars so we do not need to go into it in any great detail when discussing Jupiter.

Jupiter moves across the sky in a very predictable pattern, but every now and then it reverses direction in the sky, making a tiny loop against the background stars – this is Jupiter in retrograde.
The following diagram shows that, as Jupiter is overtaken by the Earth, its apparent motion across the sky appears to describe a loop as its direction changes from prograde to retrograde and then back to prograde again.

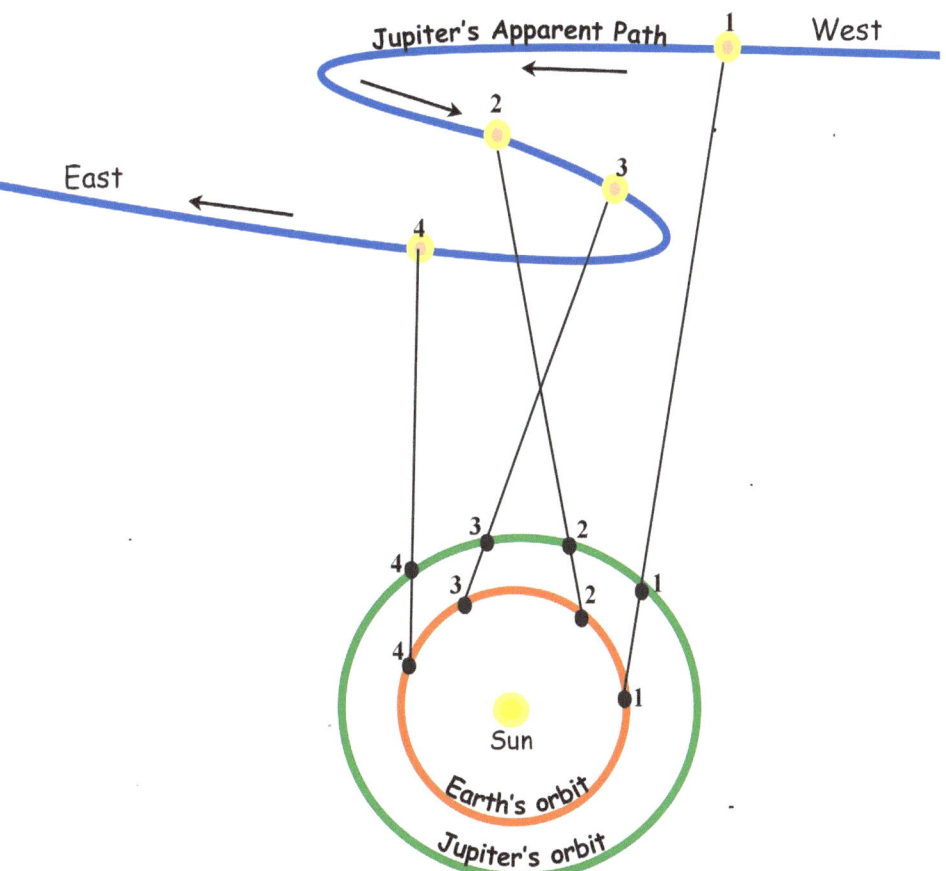

Jupiter's Retrograde Loop

At position 1, it appears to be moving from west to east in prograde motion. At positions 2 and 3, its direction appears to have changed from prograde to retrograde so that it is now moving from east to west. At position 4, it appears to have resumed prograde motion as it moves from west to east again.

Note. Sky maps can be very confusing because they are not drawn in the conventional way with east on the right and west on the left. They are drawn as they would appear in the sky if we were lying down with our legs pointing to the south and looking upwards so that east would be on our left and west on our right.

Jupiter's retrograde periods last for 4 months and are then followed by periods of nine months of prograde motion before going retrograde again. So the time from the beginning of one retrograde movement to the beginning of the next is approximately 13 months.

The relatively slow movement of Jupiter across the sky makes it very easy for the navigator to locate. Its orbit takes **11.85 earth years to complete and during that time,** it appears to move from one constellation to another every 13 months describing its **retrograde loop** as it pauses in each one. Its path leads it through Leo, Virgo Libra, Scorpius, Sagittarius, Capricorn, Aquarius, Pisces, Aries, Taurus, Gemini, Cancer and then back to Leo to begin the sequence all over again. This predictable path across the sky together with the fact that Jupiter is the fourth brightest celestial body in the sky explains why it is such an important navigational planet.

Saturn.

Named after the Roman god of agriculture, Saturn has similar characteristics to the other planets that we have discussed. It is the sixth planet from the Sun and is the second largest object in the solar system following Jupiter. Like Jupiter it is one of the gas giants and is composed mainly of hydrogen. It is the fifth brightest object in the solar system after the Sun, Moon, Venus and Jupiter and is classified as a navigational planet.

Like Jupiter, Saturn disappears behind the Sun when it is in conjunction with the Earth; however, unlike Jupiter, it does not become a 'morning or evening star' as it is not visible for about two weeks either side of conjunction.

The distance of Saturn from the Sun is constantly changing because, like the other planets, its orbit is elliptical; the mean distance is 1.4 billion km.   It travels a distance of 8.4 billion km at a speed of 9.69 km/s and takes 354 Earth months (29.5 Earth years) to complete each orbit.

Like Mars and Jupiter, Saturn's apparent motion changes between prograde and retrograde. Each of its retrograde loops lasts for an average of 140 days and the average period between each phase is an average of approximately 12.4 months.

Like Jupiter, Saturn follows a predictable path across the sky which helps us to locate it. During its 29.5 year orbit of the Sun, it moves slowly in an easterly direction through the constellations of the Zodiac describing its retrograde loops at intervals of roughly 12 Earth months as it goes. It passes Scorpius, Ophicius, Sagittarius, Capricorn, Aquarius, Pisces, Aries, Taurus, Gemini, Cancer, Virgo and Libra spending 2 to 3 years in each constellation until it eventually returns to Scorpius and begins the cycle all over again.

Saturn is an easy planet for a navigator to find; not only is it is the fifth brightest object in the sky, we always know just where to look for it as it slowly follows the path described above.

Chapter 5
Stars and Constellations

"Know The Stars And You Will Always Have A Compass".   (Michael Punk. 2002. The Revenant).

The focus of this chapter is to demonstrate ways of locating stars, specifically navigational stars. It is not intended to be an astrological tour of the constellations of the zodiac; however, the order of the signs of the zodiac can be very useful in astro navigation because it tells us the relative position of any zodiac constellation with respect to the others. For example, we know that Cancer's position in the zodiac falls between Leo and Gemini and as the following diagram demonstrates, Cancer can easily be found nestling on the ecliptic between those constellations with Gemini lying to the west and Leo to the east   From this, we can see that knowing the position of one zodiac constellation in the sky can help us to locate those adjacent to it along the path of the ecliptic

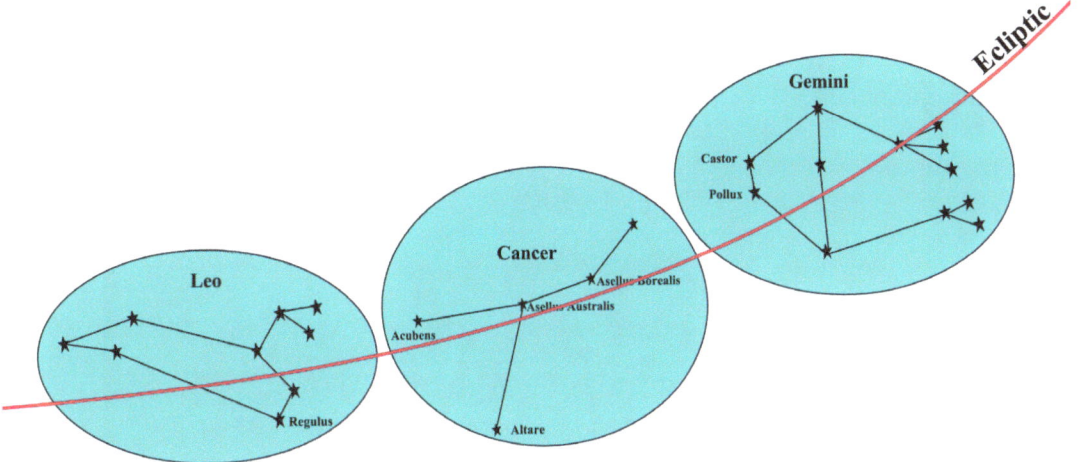

Of all the stars that are visible in the sky, only 59 are considered to be bright enough to be seen at nautical twilight when star sights are usually taken; these are known as the **Navigational stars.** As a general rule we will concentrate on the constellations that contain navigational stars; however, we will also look at some constellations that have special interest.

Have you ever wondered why the stars seem to rise earlier from night to night and why the stars that we can see change from season to season?

There are two separate reasons for these phenomena, Rotation and Revolution.

Rotation.   The Earth rotates from west to east about its axis of rotation which is a line joining the celestial poles and if this axis is produced far enough, it will cut the celestial sphere at a point marked by the North Star (Polaris) as shown in the diagram. Facing north from the Earth, the Pole Star appears stationary, and the other stars appear to rotate from east to west around the Pole Star although in fact the positions of the stars are fixed and it is the Earth which is rotating from west to east.

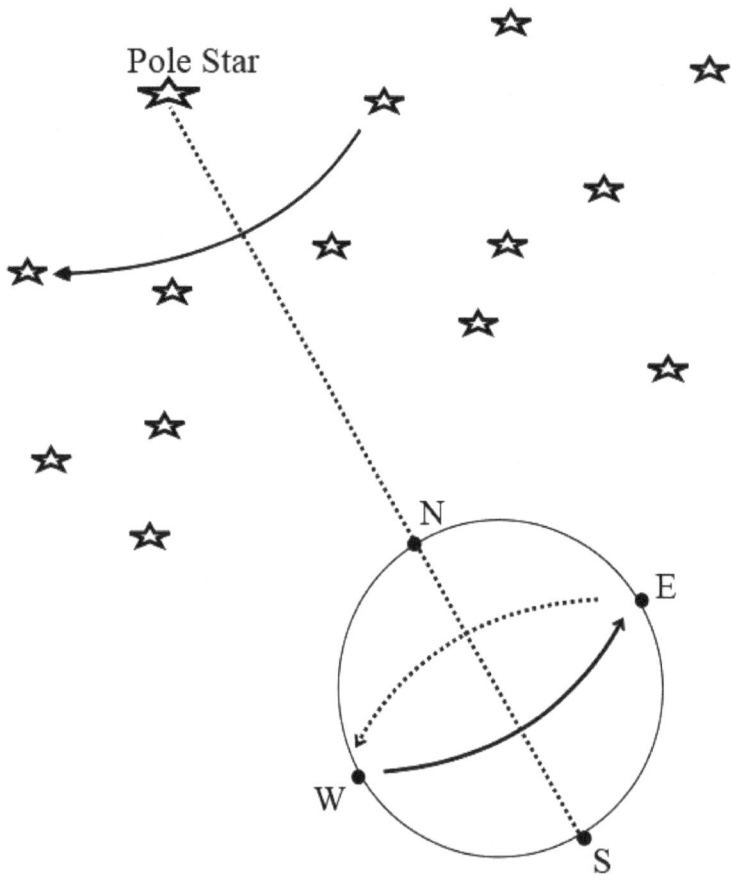

The time taken for a star to complete a circuit around the Pole Star is called a star's day or sidereal day. If the sidereal day were to be exactly 24 hours, as is the Mean Solar Day, then the stars would rise and set at the same times every day. However, the Earth completes each rotation about its axis in 23 hours, 56 minutes and 4 seconds so the stars will take the same amount of time to circuit the Pole Star and that is the length of the sidereal day. Therefore, if a star rises in the east at a certain time on a certain day, it will next do so 23 hours, 56 minutes and 4 seconds later. In other words, the star in question will rise 3 minutes and 56 seconds earlier each day (usually rounded off to 4 minutes).

For example, Say that Arcturus (the brightest star in the northern celestial hemisphere) rises at 18.00 mean time on a certain day, we know that it will rise again 23 hours and 56 minutes later so we can easily calculate that it will rise at 17.56 mean time the next day (4 minutes earlier).

Revolution.   In the diagram below we see the Earth as it orbits the Sun or to put it another way, we see it as it revolves around the Sun.

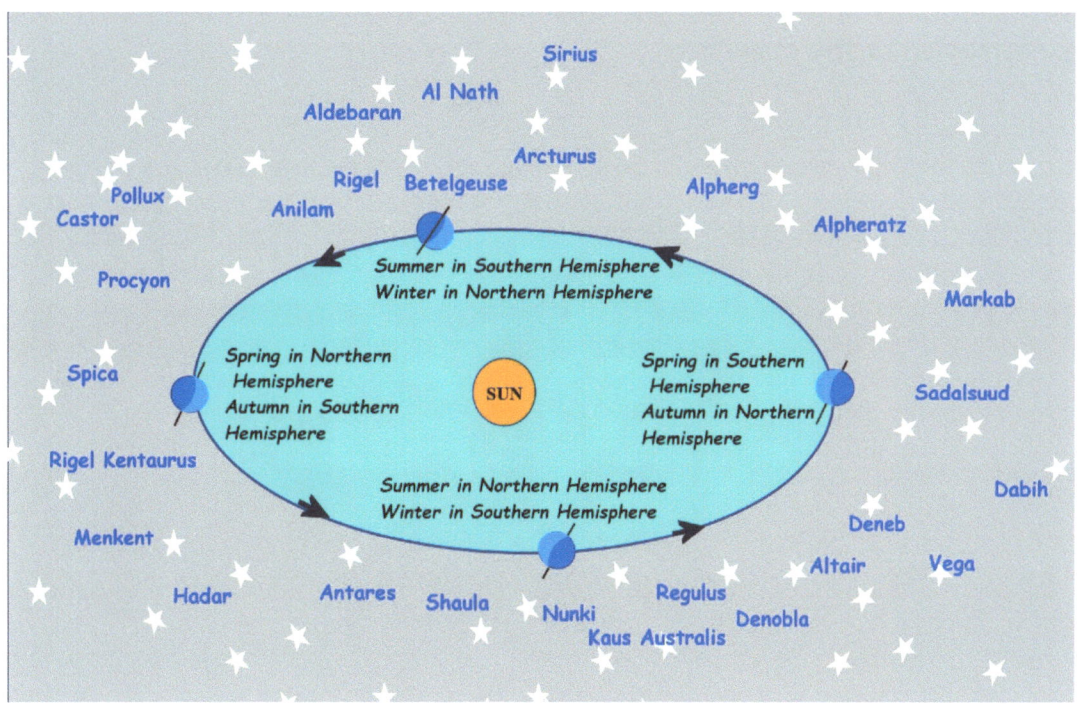

The positions of some of the more well known stars in relation to our Sun are also shown in the diagram and it can be seen that, as the Earth follows its orbital path, different stars will gradually come into and out of view in the night sky. For example, in the Northern Hemisphere, we will see Sirius in the night sky during the winter but we won't see it during the summer nights. Its not that it's in a different place, its just that it is now on our daylight side.

So, in the Northern Hemisphere, we have our Winter Stars such as Aldebaran, Rigel and Betelgeuse and we have our Summer Stars such as Nunki and Kaus Australis; of course, it is the other way round for the Southern Hemisphere.

Circumpolar Stars. A circumpolar star is one that, from a given latitude on Earth, never rises or sets because it is always above the horizon. Whether or not a star is circumpolar depends on two things, the star's declination and the observer's latitude. If the angular distance of a star from the nearest pole is less than the latitude of the observer, then that star will be circumpolar to the observer. (The angular distance of a star is the complement of its declination so if its declination is S57° then its angular distance from the South Pole will be 33°). For example, if an observer's latitude is 51°N and a star's declination is N66° then the angular distance of the star will be 24° which is less than the latitude and so it will be circumpolar to the observer.

Circumpolar Constellations In The Northern Hemisphere

Ursa Major. The best known and most easily recognizable constellation in the northern hemisphere is Ursa Major which is also known by various names such as the Great Bear, the Big Dipper and the Plough.

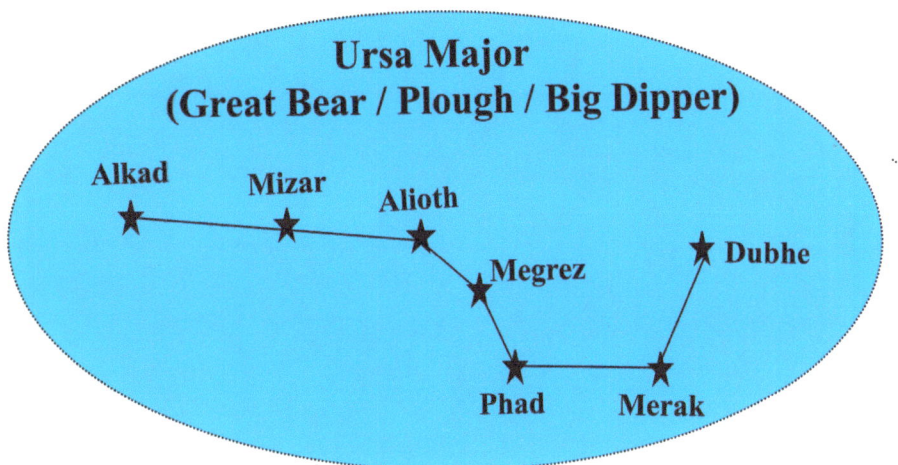

Ursa Major contains 3 navigational stars named Dubhe, Alioth and Alkaid; it is circumpolar from about 42° North (the angular distance of Alkaid is 41°).

Ursa Minor. The Little Bear (also known as the Little Dipper). Ursa Minor contains the star **Polaris** which has a declination of 89°16'N. almost coinciding with the Celestial North Pole and for this reason it is also known by various names including **Pole Star, North Star, Lodestar and the Guiding Star.**

Polaris is only the 45[th.] brightest star in the sky; however, it has always played an important role in navigation; not only because it indicates the direction of north but also because it is useful for position fixing in the north polar-regions. Polaris is circumpolar in the Northern Hemisphere from just above the Equator while Kochab, the southernmost star in Ursa Minor which has a declination of N74° is circumpolar from latitudes north of 16°N.

Finding the Pole Star. Ursa Major contains a reference line known as the line of pointers. The line joining Merak to Dubhe, when extended will point to Polaris in the constellation Ursa Minor as illustrated in the diagram above.

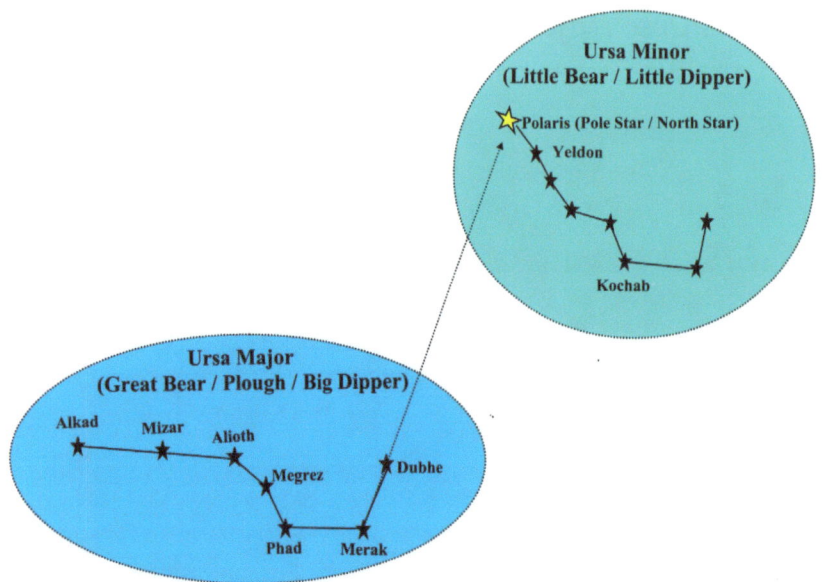

Ursa Major and Ursa Minor are associated with several mythological stories. In one such story, the two nymphs who had nursed Zeus as an infant were sent into the sky by him to form constellations. The nymph Adrasteia became the constellation Ursa Major while the other, Ida became Ursa Minor.

Cassiopeia. Cassiopeia is quite an easy constellation to find because of its 'W' shape which sometimes appears to be hanging upside-down as it revolves around the Pole Star. The brightest star in Cassiopeia is Alpha Cassiopeia otherwise known as Schedir which is a navigational star and is circumpolar above 32°North.

The constellation Cassiopeia is associated with Queen Cassiopeia in Greek mythology. In punishment for her vanity, she was made to sacrifice her daughter Andromeda and as further punishment, she was sent into the sky to circle the North Pole forever.

How to find Cassiopeia. Cassiopeia can be located along a line of reference from the Pole Star at an angle of 135° to the line of pointers in Ursa Major, as the diagram below shows. As Ursa Major revolves around the Pole Star, so do the five stars of Cassiopeia but Segin always keeps its position 135° from the line of pointers. The angular distance of star Segin from the Pole Star is 30° or roughly one and a half hand-spans.

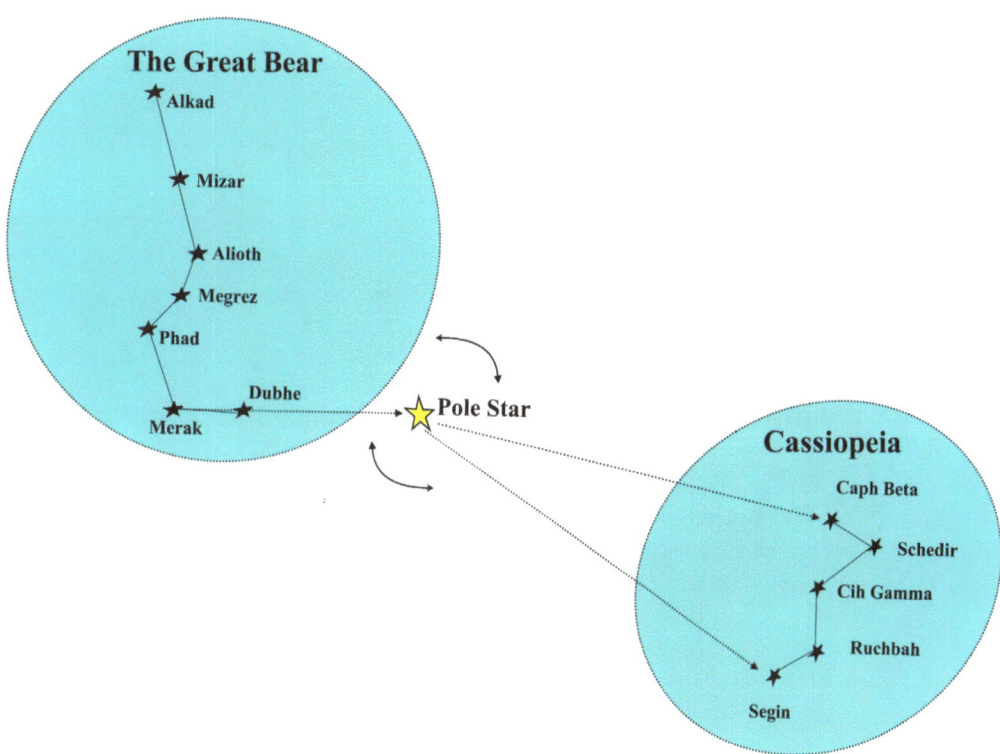

Perseus. The constellation Perseus is associated with the Greek mythological hero Perseus who, on the orders of King Polydectes, slayed the Gorgon Medusa who had the power to turn people to stone. Polydectes had hoped that Perseus would not return and became hostile when he did. Perseus was so angered by this that he took out the head of Medusa and turned Polydectes to stone. The wife of Perseus was called Andromeda and the constellation that represents her lies side by side with the one that represents him.

Even though Perseus's stars are bright relative to other constellations, Mirphak which is circumpolar in the Northern Hemisphere above 41°N, is its only navigational star. It should be pointed out that although Mirphak is circumpolar, the whole constellation Perseus is visible throughout the northern hemisphere and in the northern areas of the southern hemisphere during spring and early summer.

If a line is drawn from Navi to Ruchbah in Cassiopeia, it will point almost directly towards the star Mirphak of the constellation Perseus at about one hand-span as shown in the diagram below.

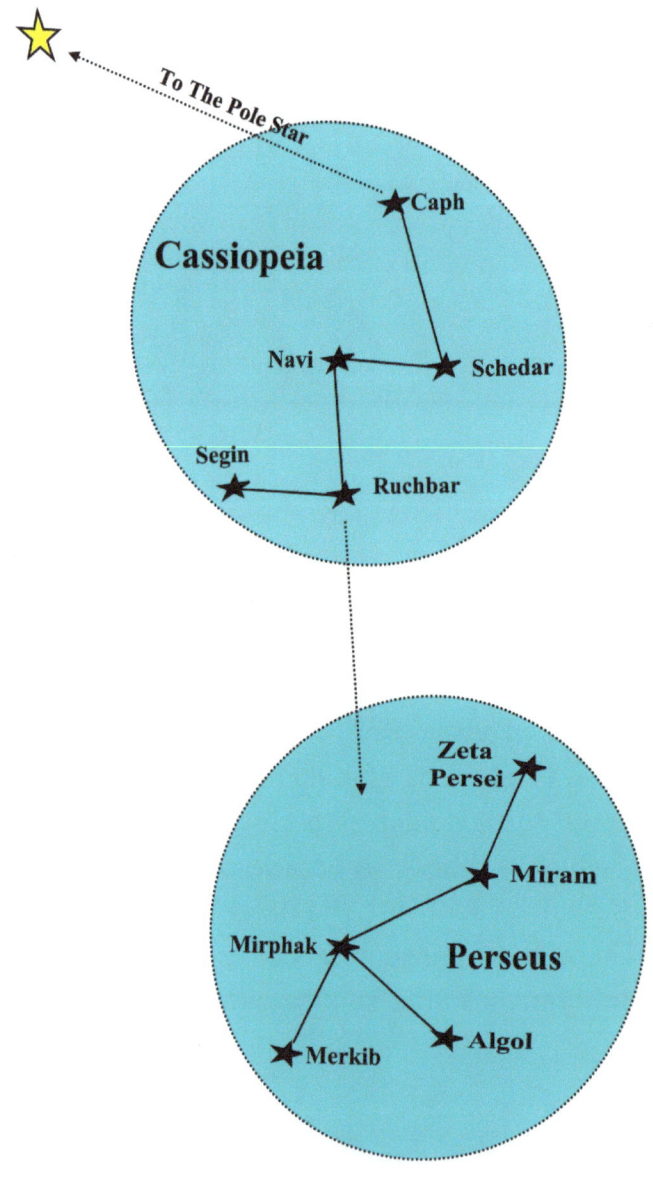

Circumpolar Constellations In The Southern Hemisphere.

Centaurus.   Centaurus is the ninth largest of all constellations and extends from about 30° south to 61° south. It is one of the brightest constellations in the sky and contains three navigation stars: Hadar, Menkent and Rigil Kentaurus. Hadar and Rigil Kentaurus are circumpolar in the Southern Hemisphere from about 31° south while Menkent, the most northern of the three, is circumpolar from 55° south.

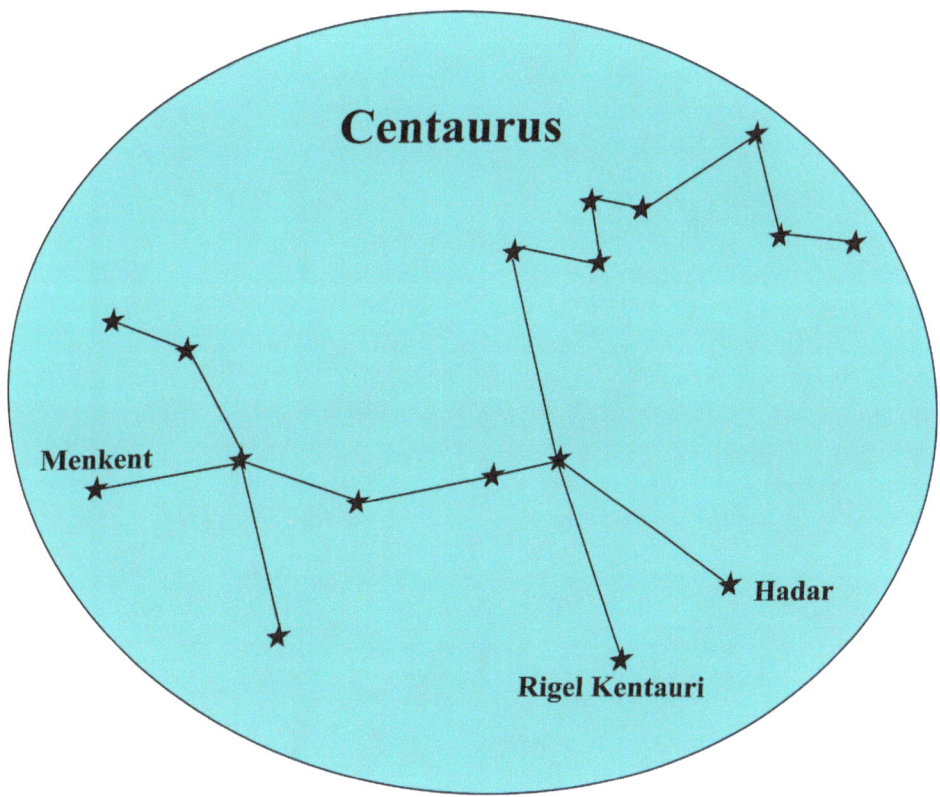

Centaurus is associated with several mythological tales and is depicted as a centaur, half man and half horse. In Ancient Greek mythology, the constellation is associated with Chiron, the centaur who mentored many of the Greek heroes including Theseus, Jason and Heracles.

Finding Centaurus. From February to late June when Boötes and Virgo are in the sky, they can be used to help find Centaurus. If we take a line from Arcturus in the constellation Boötes to Spica in Virgo and extend that line by another 55°, it will point to the constellation Centaurus as shown in the diagram below.

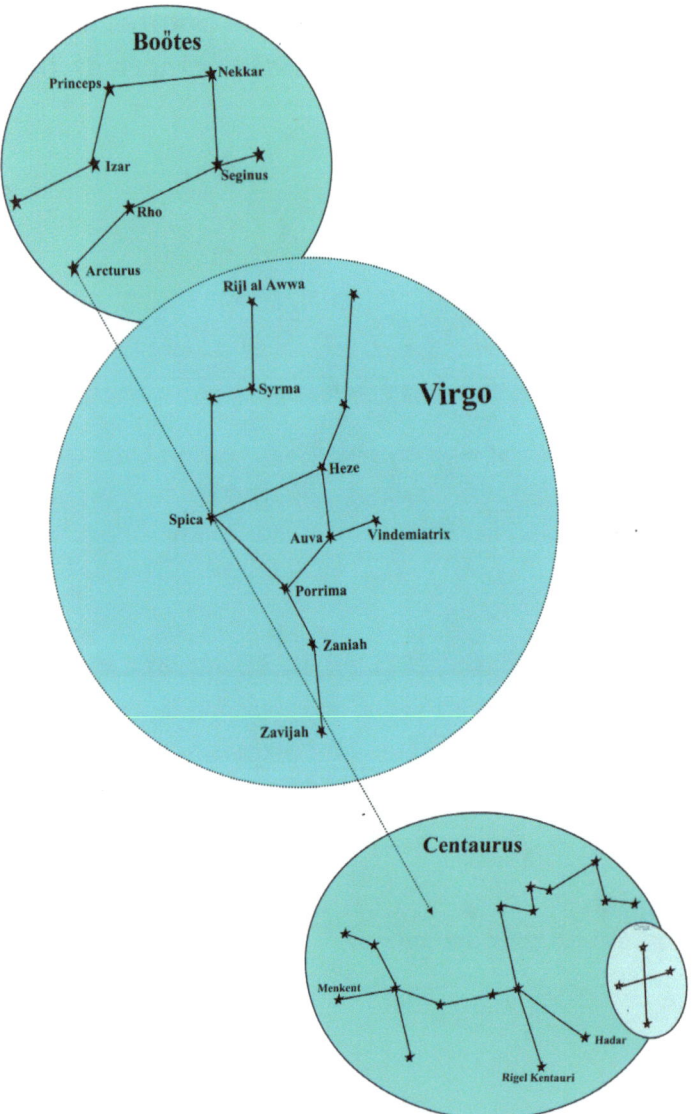

Crux, The Southern Cross Crux is one of the best known constellations in the southern hemisphere, and is easily recognizable for the cross-shaped asterism named the 'Southern Cross'. Crux (Latin for cross) is one of the smallest constellations in the sky but also one of the brightest which makes it useful for astro navigation.

The cross has four main stars: alpha, beta, gamma and delta which mark the tips of the cross.

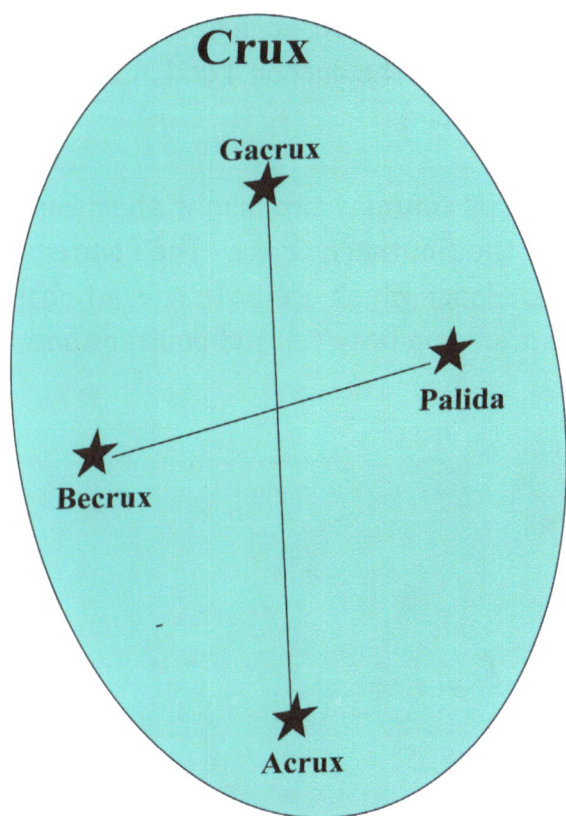

Alpha Crucis is also known as Acrux (a contraction of 'alpha' and 'Crux'). This is the brightest star in the constellation Crux; it is a navigational star and is circumpolar south of 28° south .

Beta Crucis, also known as Mimosa or Becrux, is the second brightest star in the constellation but is not a navigational star; it is circumpolar south of 40° south .

Gamma Crucis or Gacrux, is the third brightest star in Crux; it is a navigational star and is circumpolar below 32° south .

Delta Crucis or Palida is the fourth star and has variable levels of brightness making it unsuitable as a navigational star; like Beta Crucis, it is circumpolar south of 40° south.

Mythology.   Crux is associated with the mythology of several cultures. Some Aboriginal cultures saw it the head of the 'Emu in the sky' while others saw it as representing the sky deity Mirrabooka and indeed that is the name that they gave to what we know as the 'Southern Cross'. In New Zealand, the Maori called the cross Te Punga ('the anchor') and in South America, the Incas called it Chakana (the stair).

Finding Crux.   The constellation Centaurus contains two bright stars which make excellent pointers to help us find the Southern Cross. The Pointers as they are known, are Rigil Kentaurus and Hadar which are both navigational stars. The next diagram shows the relationship between the pointers and the Southern Cross.

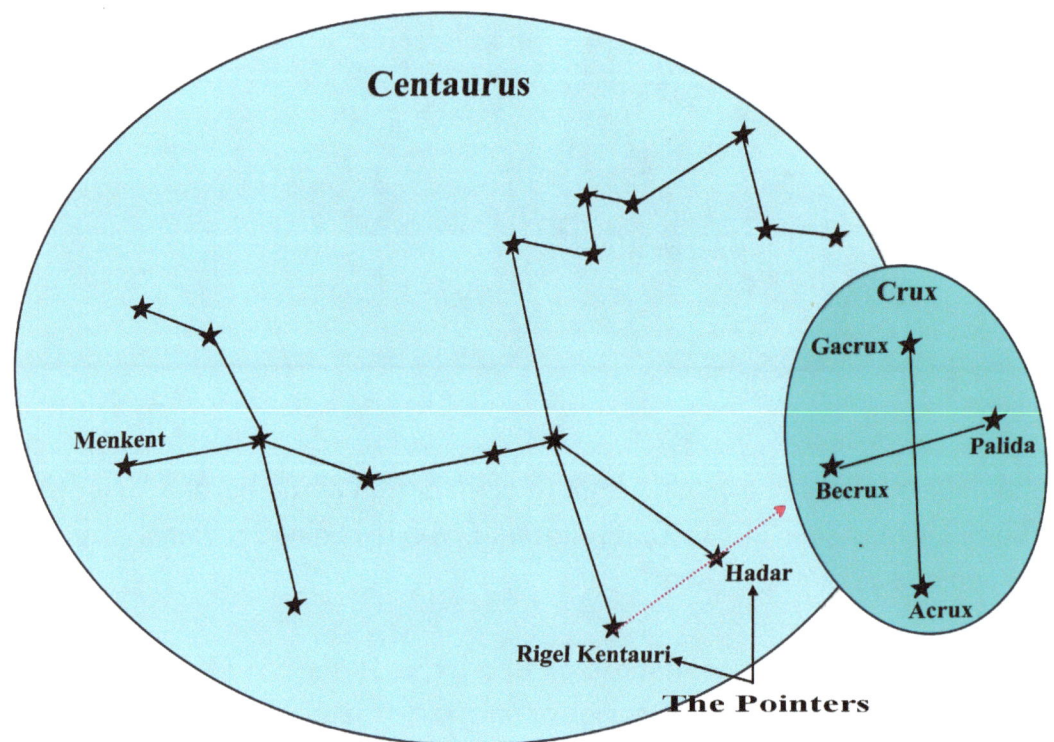

The constellation Centaurus appears to envelope Crux and indeed the ancient Greeks considered them to part and parcel of the same constellation.

How to find the direction of south by using the Southern

Cross. There are several methods of doing this but the simplest is as follows: Make an imaginary line between Gacrux and Acrux then extend this line from Acrux (the brightest star) for 4.5 times the length of the Southern Cross, as shown in the diagram below. This will take you to the position of the South Celestial Pole in the sky. From the South Celestial Pole, drop a line down to the horizon. Where this line touches the horizon is the direction of south.

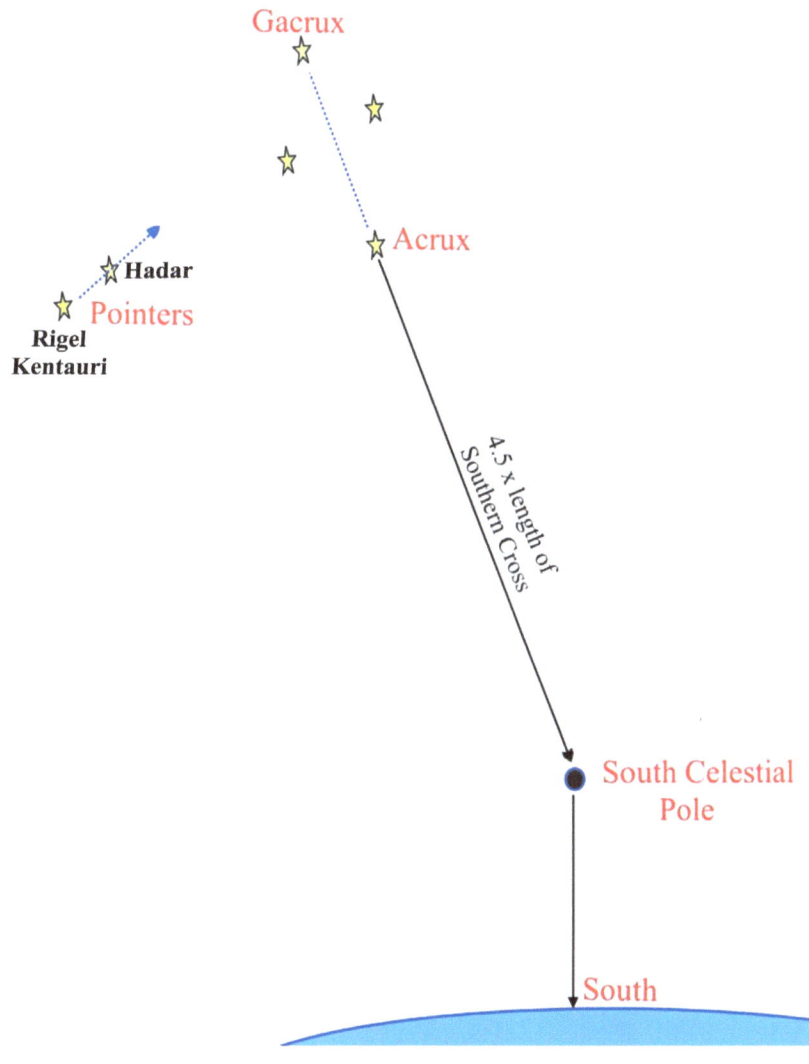

Carina

Carina lies over the Southern Hemisphere and can be seen by observers located between 20°N and 90°S. It is the 34th biggest constellation and it contains two navigation stars, Canopus and Miaplacidus. Canopus is circumpolar below 38°S; it is second brightest star in the night sky and can be seen to the south of Sirius, the brightest star which lies in Canis Major. Miaplacidus is circumpolar below 21°S; it is the second brightest star in Carina but only the 29th. brightest in the night sky. Foramen (scientific name Eta Carinae) is circumpolar south of 30°S. It is a variable star and was once the brightest star in the sky but currently its apparent magnitude is around 4.5.

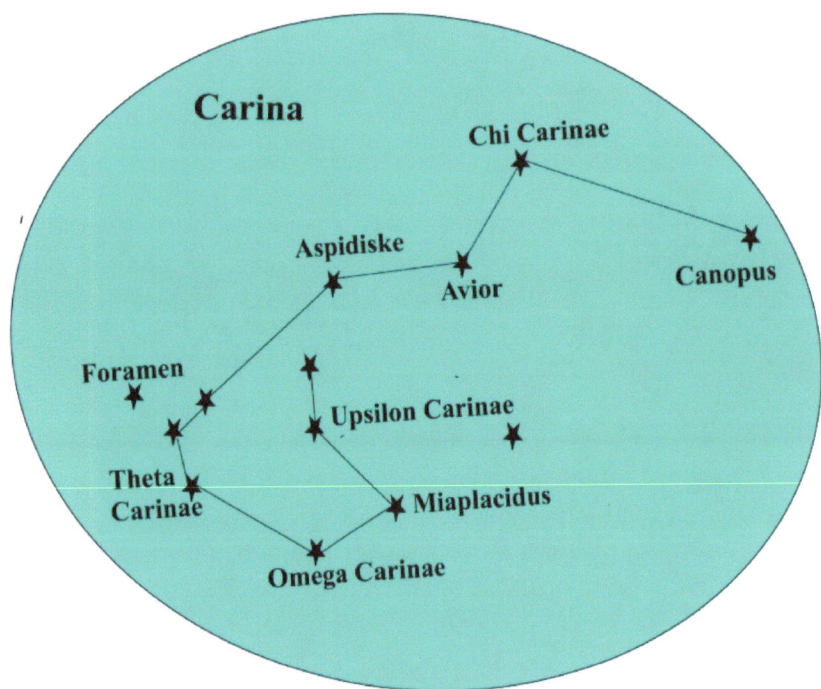

Carina was once part of a larger constellation Argo Navis which represented the ship on which Jason and the Argonauts sailed to find the Golden Fleece. In that constellation, Carina represented the hull of the ship and the star Canopus marked the blade on one of the ship's steering oars.

In 1763, Argo Nevis was divided into three smaller constellations: Carina, (the hull), Puppis (the stern) and Vela (the sails).

Vela. Vela represented the sails of the Argo Navis. Its brightest star, Suhail has an apparent magnitude of 1.7 and is a navigation star; it is circumpolar below 43°S

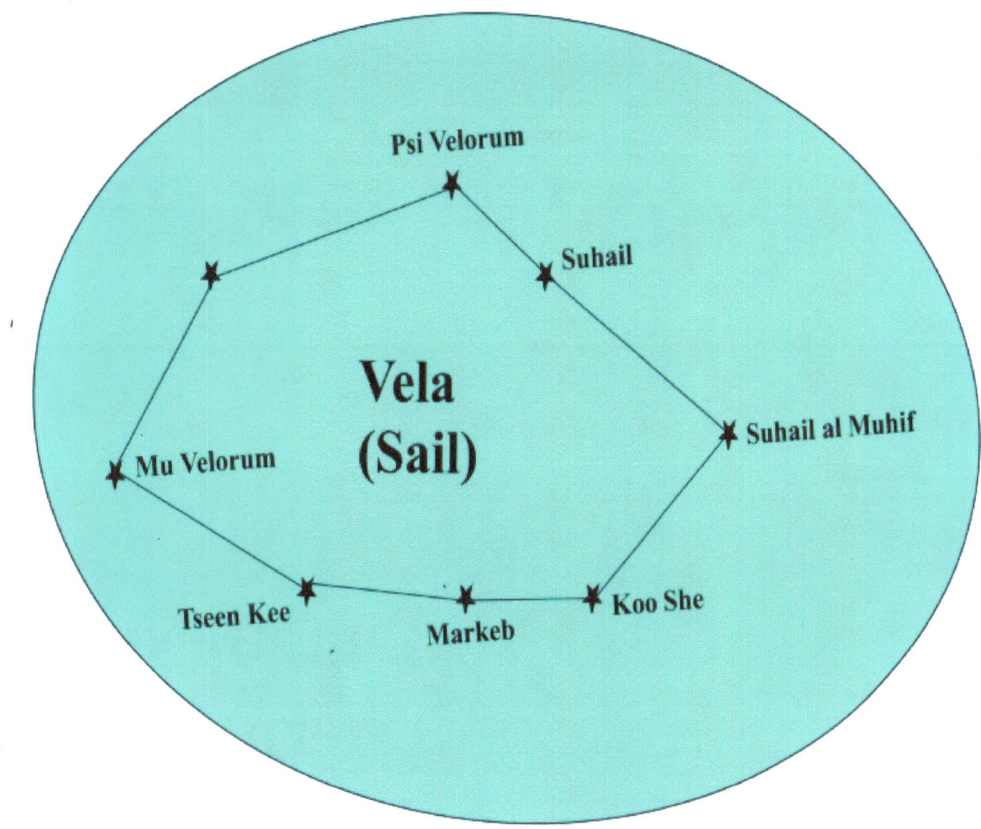

The star Markab, (declination 55°S) should not be confused with the star Markab that lies in the constellation Pegasus (15°N). Vela's Markab is not considered to be a navigation star even though it is slightly brighter than the Markab of Pegasus which is a navigation star. With a declination of 55°S Vela's Markab is circumpolar below 36°S.

Koo She, which is Chinese for 'bow and arrows', is the second brightest star in the constellation but is not considered to be a navigation star even though it has a magnitude of 1.96. The declination of Koo She is 54°S and it is therefore circumpolar from 36°S.

Puppis.

Puppis was said to be the stern of the Argo Navis but was sometimes known as the 'poop deck'. It has only two named stars and neither of these are classified as navigation stars.

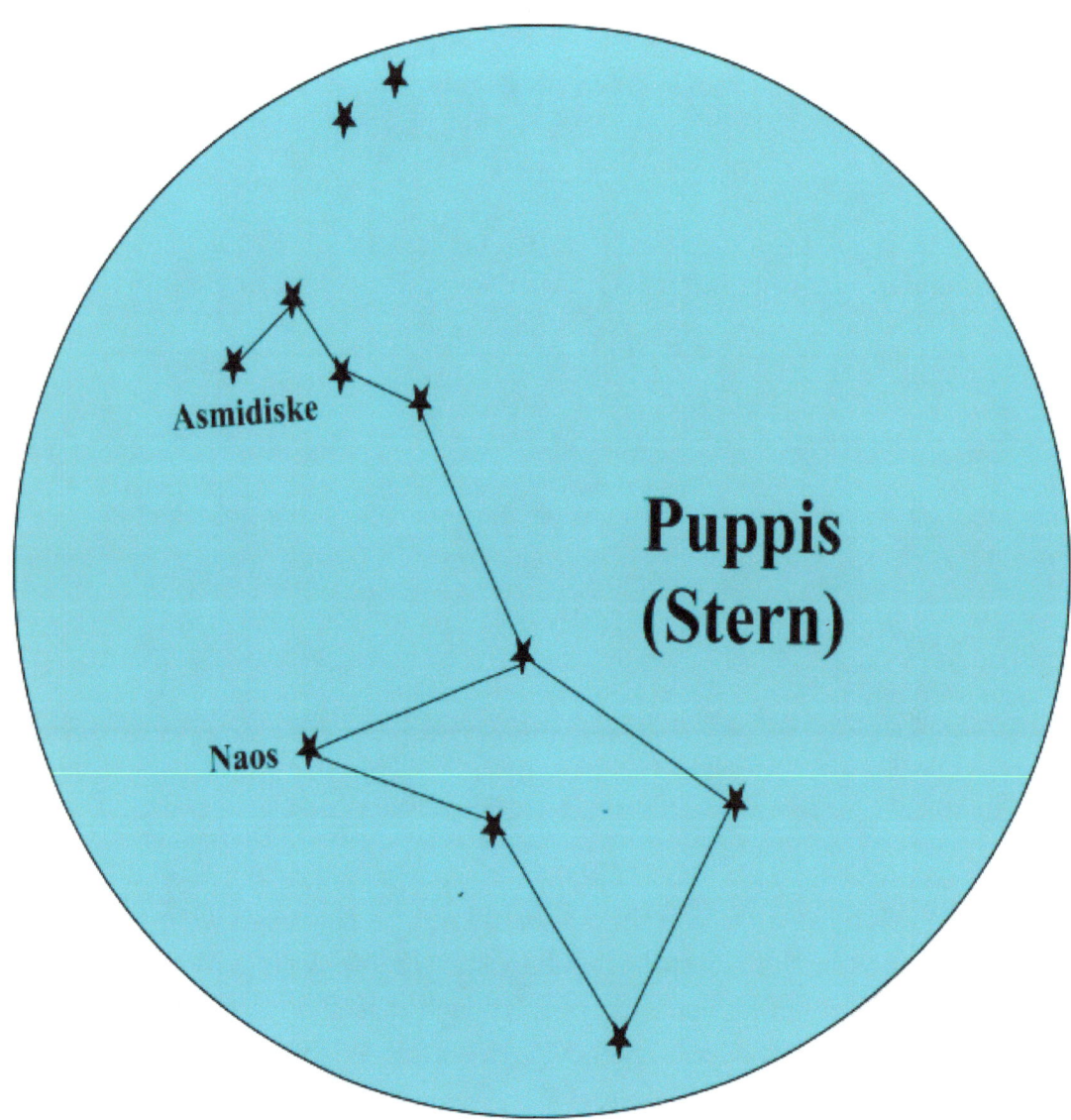

The star Naos has a declination of 40°S and is therefore circumpolar below 51°S.

The star Asmisdiske with a declination of 25°S is circumpolar below 66°S.

The diagram below shows how the three new constellations once fitted together to form the ship Argo Navis and if we continue to think of them in that way, we will have a ready-made aid to locating them in the night sky.

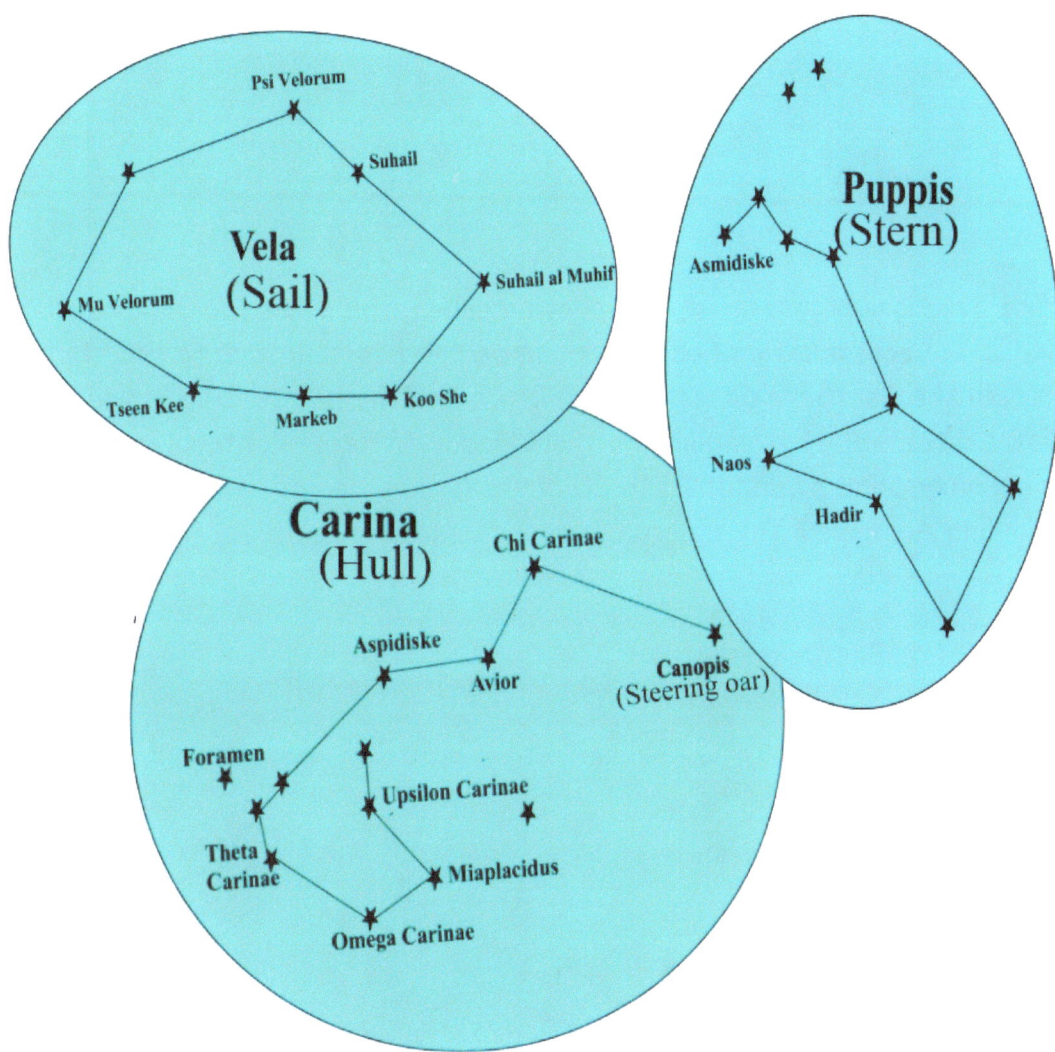

Triangulum Australe (The Southern Triangle). This is a small constellation located over the southern hemisphere. Its name is Latin for "the southern triangle", which distinguishes it from Triangulum, the Summer Triangle, in the northern sky. Its name is derived from its three brightest stars Atria, Betria and Gatria which form an almost equilateral triangle and are sometimes called the 'Three Patriarchs'. Atria is the brightest of the three and the only navigation star in the constellation. Atria and Gatria are circumpolar at latitudes greater than 21°S while Betria, the northernmost of the three, is circumpolar below 28°S.

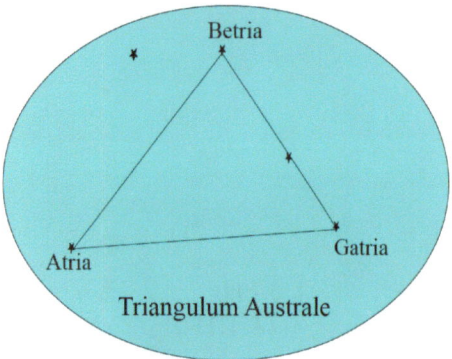

Finding Triangulum Australe. As shown earlier, the stars Rigel Kentauri and Hadar in the constellation Centaurus can be used as pointers to the star Becrux in the Southern Cross (Crux). If we line the pointers up in the opposite direction, that is pointing from Hadar to Regel Kentauri and extend that imaginary line by about 15 to 20 degrees it will point to Betria in Triangulum Australe. The diagram below demonstrates this.

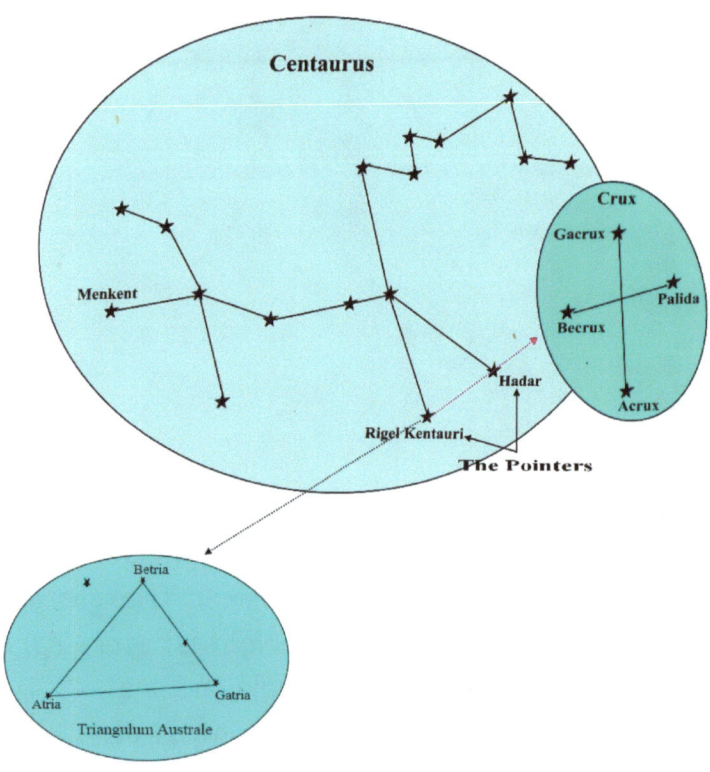

Pavo. If we draw an imaginary line between the stars Gatria and Atria in Triangulum Australae and extend that line westwards by about 20° it will point to the star Peacock, the brightest star in the constellation Pavo. Peacock, is a navigation star and is circumpolar below 34°S as are all the stars in Pavo.

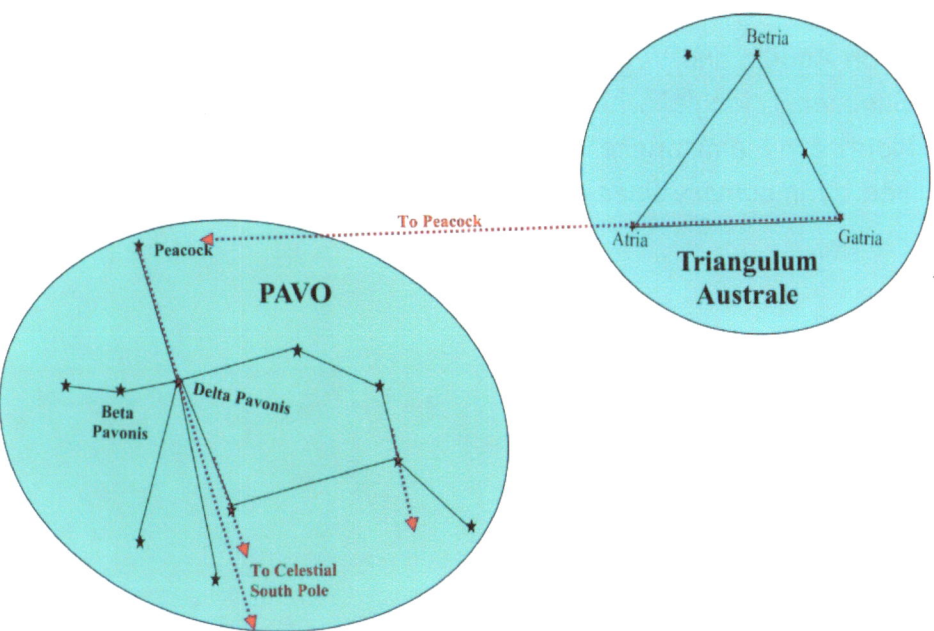

Some of the stars in Pavo form an asterism called "the Saucepan", which has been used as a navigational tool to point toward the southern celestial pole; however, because most of the stars in the saucepan are very faint, it is difficult to pick it out with the naked eye.   Another method is to imagine a line from the star Peacock to the star Delta Pavonis, the third brightest star in Pavo, and as shown in the diagram above, this line will point to the Celestial South Pole.

Summer Stars in the Northern Hemisphere (Winter Stars in the Southern Hemisphere).

Next, we look at some of the stars that we see in the Northern Hemisphere's Summer night sky.

The Summer Triangle. The stars Deneb in the constellation Cygnus, Altair in the constellation Aquilla and Vega in Lyra form an astronomical asterism known as the 'Summer Triangle' which can be seen during summer and autumn in the Northern Hemisphere. The diagram below shows how the triangle is formed by imaginary lines drawn between those stars.

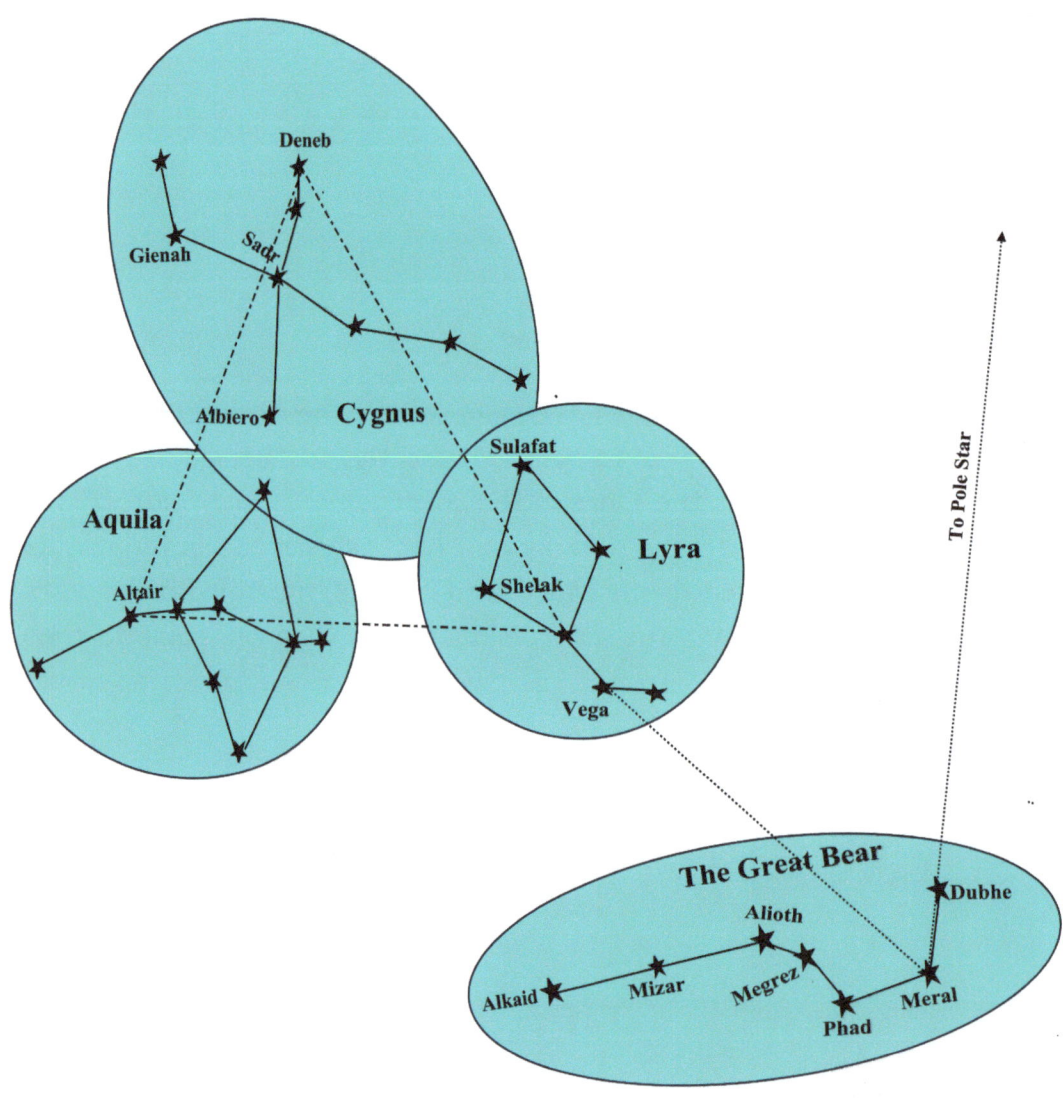

Finding the Summer Triangle.   In the diagram above, if we join the star Meral in the constellation Great Bear (Ursa Major) to a point midway between the stars Alioth and Dubhe also in the Great Bear and then extend this line, it will point to Vega, the brightest star in the constellation Lyra. In this way, The Great Bear gives us a sign post to the Summer Triangle.

Lyra The Harp. Lyra is visible between latitudes 90°N and 40°S. From a navigator's point of view, it is best seen during nautical twilight in the late summer and autumn. Lyra contains Vega, which is the second brightest star in the northern hemisphere and is a navigational star.

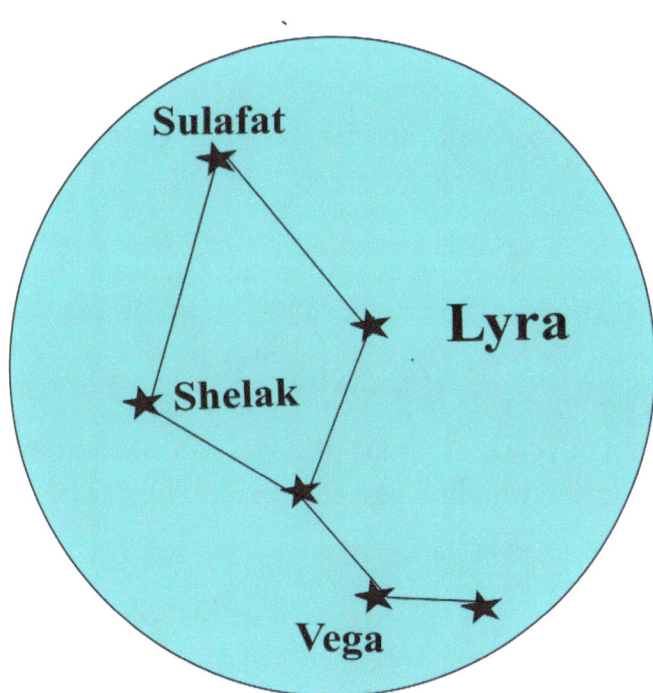

In Greek Mythology, when Orpheus died, he dropped his lyre into a river from where it was retrieved by an eagle sent by Zeus. Zeus then sent both the lyre and the eagle into the sky as the constellations Lyra and Aquila.

Aquila, The Eagle. Aquila is visible between latitudes 85°N and 75°S and is also best seen during nautical twilight during the late summer and autumn.

Aquila contains Altair, the 12th brightest star in the sky and also a navigational star.

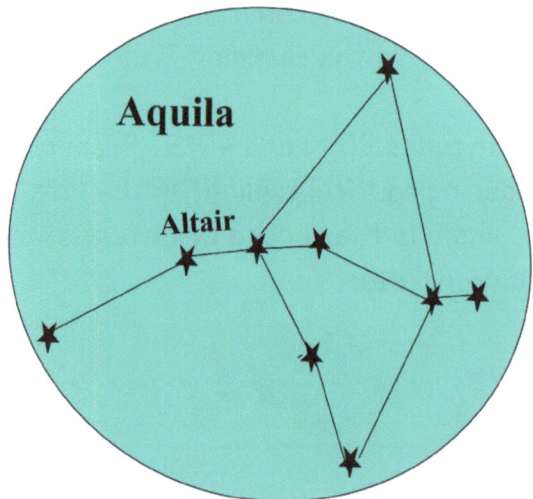

In Greek mythology, Aquila, the eagle, carried thunderbolts for Zeus and as explained above, later rescued the lyre of Orpheus from the river.

Cygnus The Swan (The Northern Cross).   The constellation Cygnus contains 6 stars, the brightest of which is Deneb, the 19th brightest star in the sky and a navigational star. As with Lyra, Cygnus is visible between latitudes 90ºN and 40ºS. For navigators it is best seen during nautical twilight in late summer and autumn. Cygnus contains an asterism formed by the brightest stars in the constellation which is named the Northern Cross.

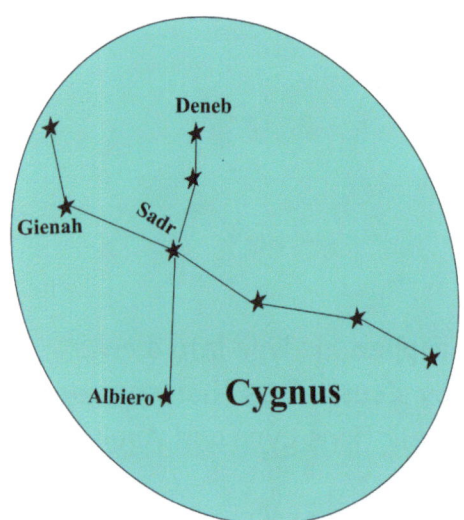

In Greek mythology, Orpheus was said to have been turned into a swan by Zeus and sent into the sky as the constellation Cygnus along with Aquila and Lyra.

Sagittarius, The Archer. Sagittarius is a large constellation lying over the southern hemisphere and is visible between latitudes 55°N. and 90°S. It contains several bright stars including two navigational stars, Nunki and Kaus Australis which are best seen during nautical twilight in August.

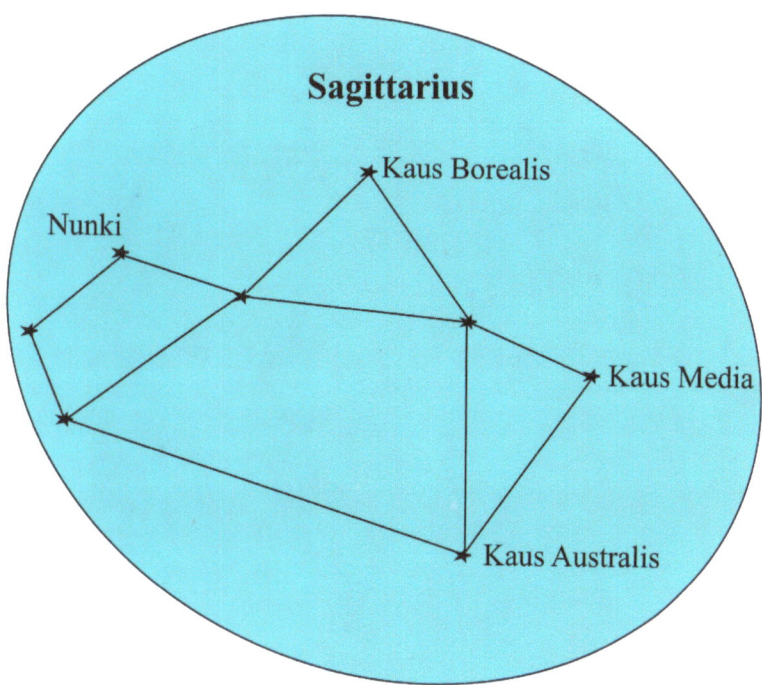

Point of Interest - The Tropic of Capricorn. Nowadays, the Sun is over the constellation Sagittarius at the Winter Solstice on 21/22 December when the Sun's declination reaches its southernmost latitude of 23.4° south. However, in ancient Greek times, the Sun passed through the constellation Capricornus at this time hence the reason for naming the latitude 23.4° south the Tropic of Capricorn.
The constellation Capricornus is not included in this chapter because it does not contain any navigational stars.

Finding Sagittarius. The Summer Triangle provides a useful pointer to Sagittarius. If we draw an imaginary line from the star Deneb through the star Altair in the Summer Triangle and extend that line by about 20° or one hand-span, it will point to the constellation Sagittarius as the diagram below shows.

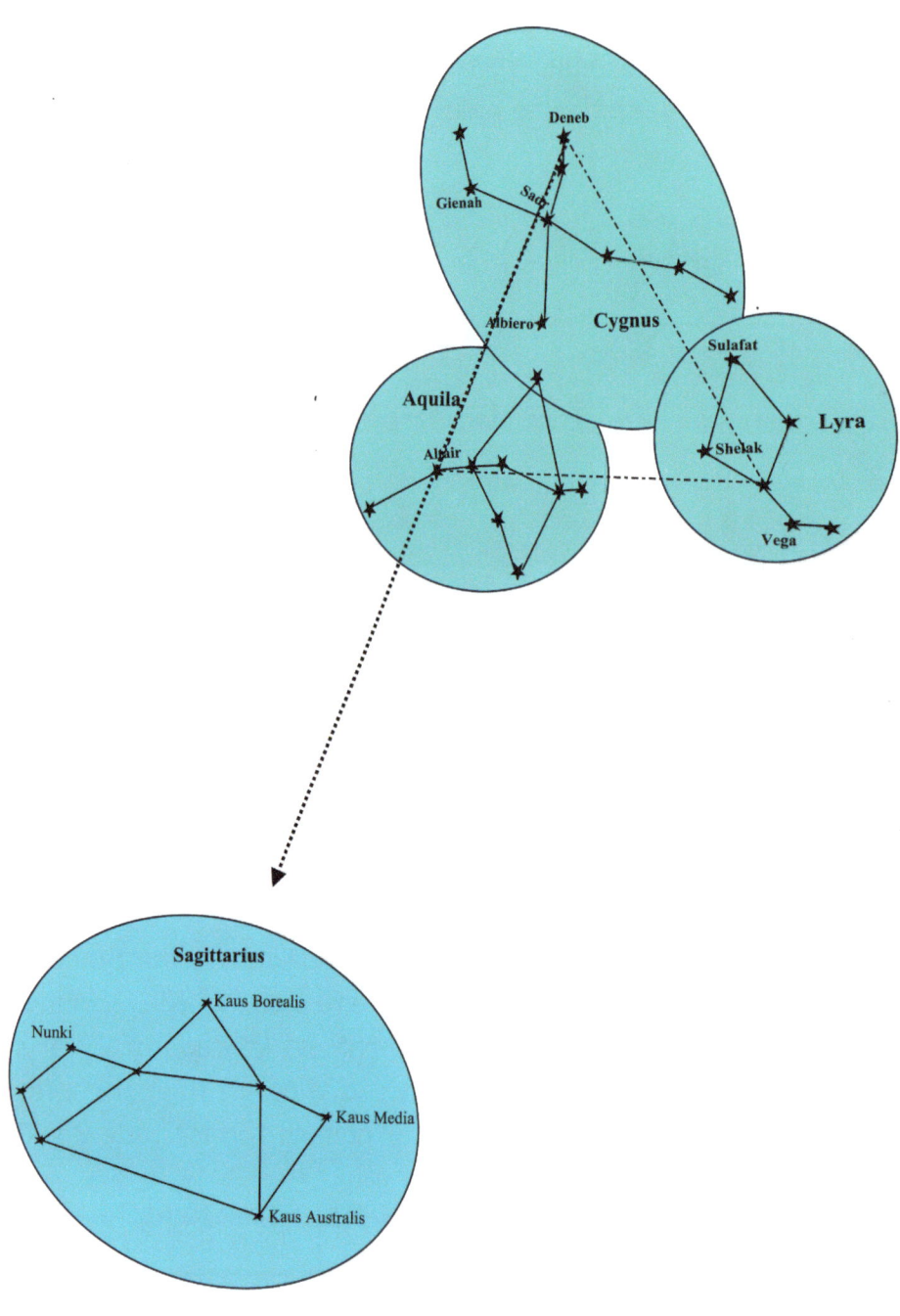

Scorpius, The Scorpion. The constellation Scorpius lies above the southern hemisphere and is visible between latitudes 40ºN and 90ºS.

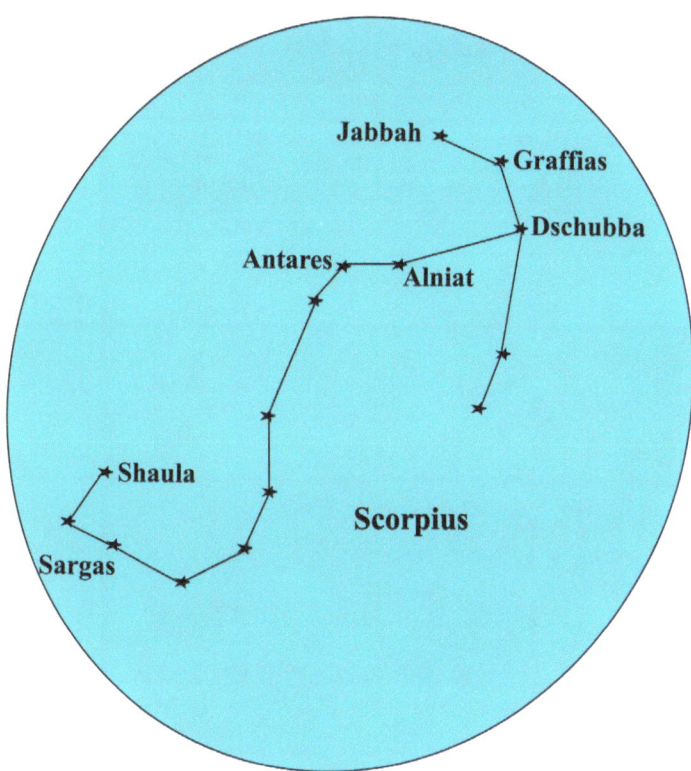

Scorpius has several bright stars which, between them, form the shape of a scorpion. The brightest star in Scorpius is Antares which is often mistaken for Mars because of its reddish orange colour. Antares is the 16th brightest star in the sky and is a navigational star. The second brightest star in Scorpius is Shaula which is said to represent the sting in the tail of the scorpion. Shaula is also a navigational star. For navigation purposes, Antares and Shaula are best seen during nautical twilight in July.

In Greek mythology, Scorpius represents the scorpion that the goddess Artemis sent to sting and kill Orion who had tried to ravish her but Sagittarius, the archer who was a centaur, half man and half horse, was sent to slay the scorpion. In the representation of the two constellations below, Sagittarius has a drawn bow with the arrow pointing to the star Antares, the heart of Scorpius.

This legend helps us to find and identify both Scorpius and Sagittarius. The line from Nunki to Kaus Media in Sagittarius represents the arrow, the head of which points to Antares in Scorpio. It also helps to remember that the orange star Kaus Media points along the line of the arrow towards the red star Antares. The bright red glow of Antares further helps us to identify Scorpius.

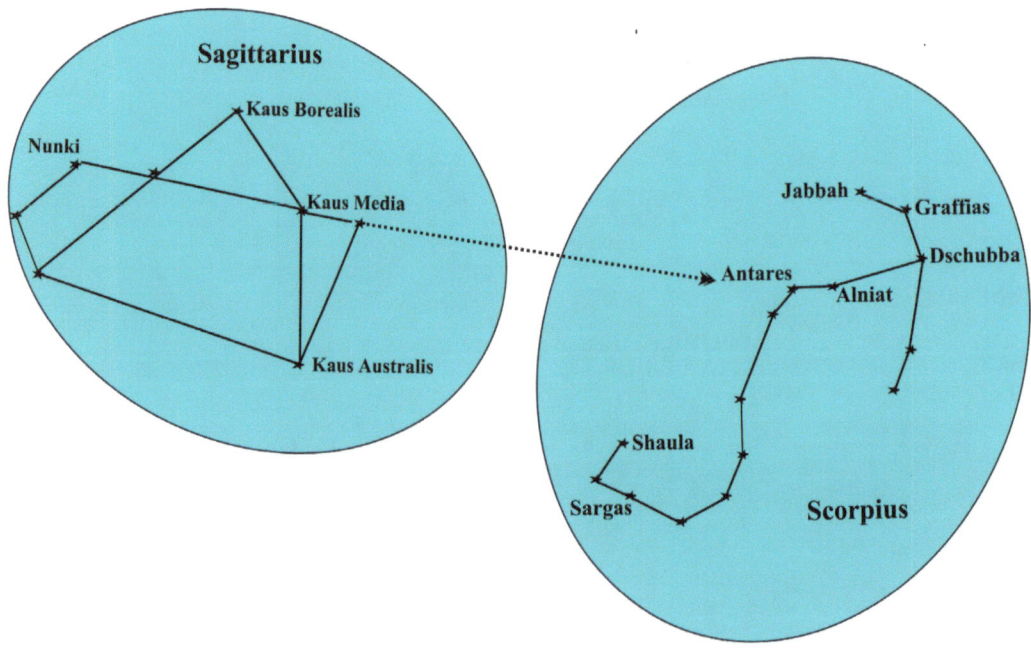

Autumn Stars in the Northern Hemisphere (Spring Stars in the Southern Hemisphere).

During the early Autumn, as the Earth continues to orbit the Sun, the last of the summer stars such as Altair, Vego, Deneb, Nunki and Kaus Australis move away to the west. Other stars take their place in the northern night sky such as Alpheratz of the constellation Andromeda, Sadalsuud of Aquaries and Markab of Pegasus.

Andromeda. The Chained Maiden.

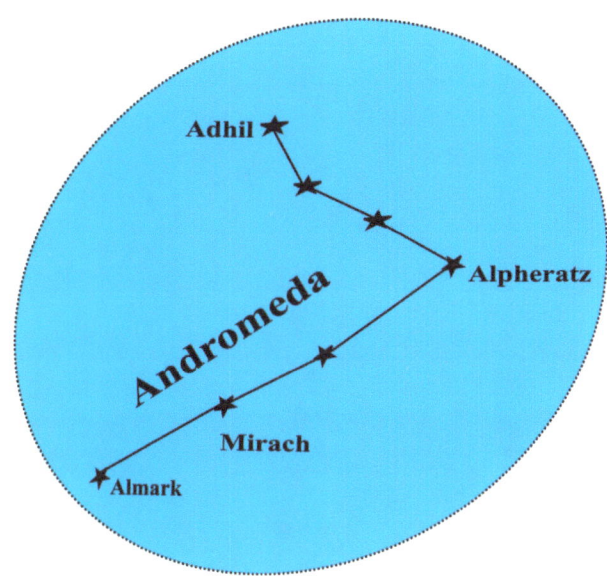

Andromeda is a constellation in the northern hemisphere and is visible between latitudes 90°N and 60°S. The brightest star in Andromeda is Alpheratz which is also included in the constellation Pegasus. **Alpheratz** is a navigational star which, for navigators, is best seen in the northern hemisphere during nautical twilight in the month of November.

The constellation Andromeda is named after Andromeda, the wife of Perseus in Greek Mythology. It is sometimes called the 'Chained Maiden' because according to legend, Andromeda was rescued by Perseus who found her chained to a rock and left as a sacrifice to the monster Cetus.

How to find Andromeda. The next diagram shows that, if a line from the Pole Star to Segin in Cassiopeia is extended by about one hand-span, it will point to the star Almark of Andromeda; alternatively, a line from the Pole Star through the star Caph Beta of Cassiopeia will point to Alpharatz in Andromeda. (Cassiopeia is one of the northern hemisphere's circumpolar constellations).

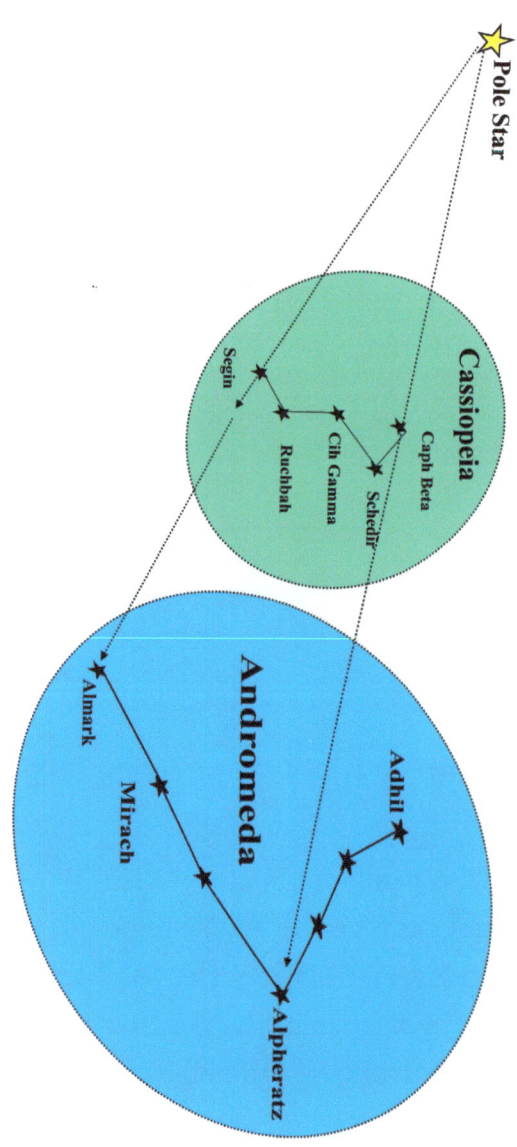

Page 74

Pegasus. The Winged Horse.

As shown in the following diagram, once we have found Andromeda, we will also have found the constellation Pegasus because the star Alpheratz in Andromeda is also included in what astronomers call the 'The Great Square of Pegasus'.

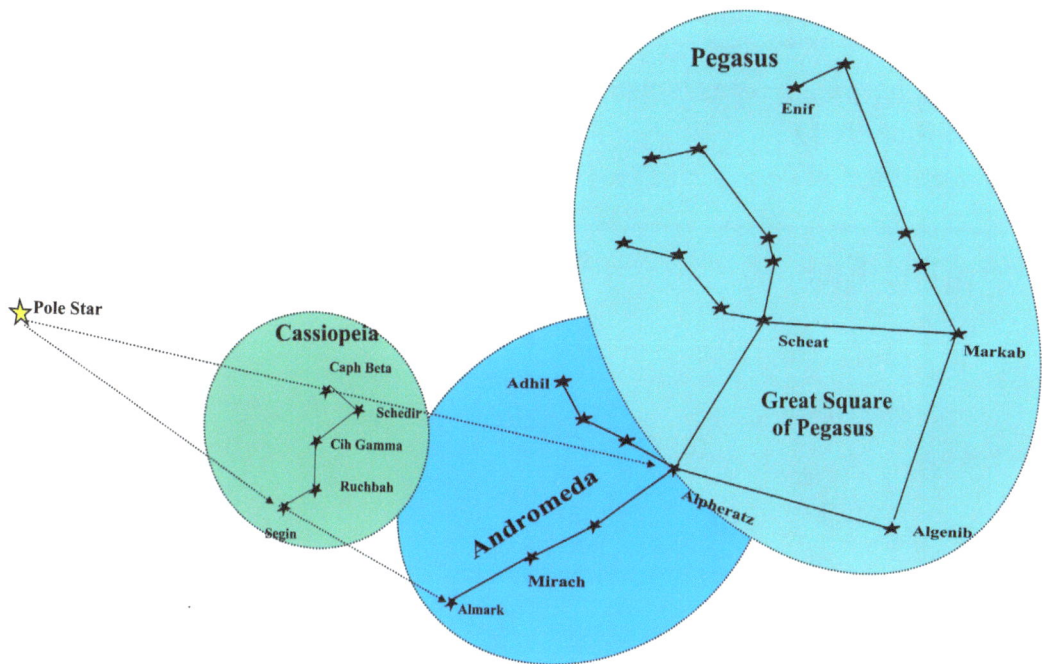

Pegasus, which is the 7th largest constellation in the sky, is in the northern hemisphere and can be seen from 90°N to 60°S. It has two navigational stars, **Enif** and **Markab** which, for navigators, are best seen during nautical twilight in the month of October.

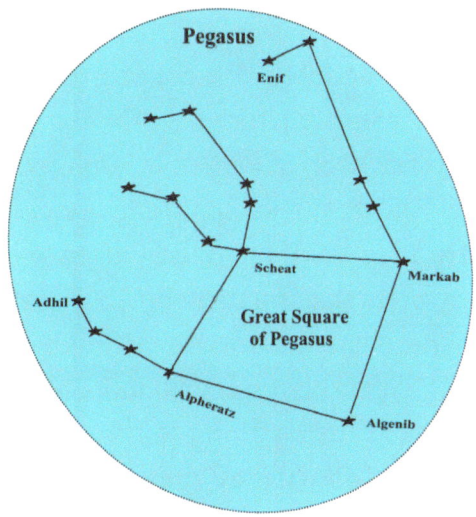

In Greek mythology, the winged horse Pegasus is said to have leaped from the body of the Gorgon Medusa after she had been slain by Perseus. The hero Bellerophon tamed the winged horse and tried to ride it to Olympus. However, Bellerophon fell from Pegasus but the horse made it to Olympus where it was kept by Zeus to carry his thunder and lightning.

The 'Square of Pegasus' is a large asterism which is said to mark the body of the winged horse. This asterism is formed by the stars Scheat, Markab, Algenib and Alpheratz.
The brightest star in Pegasus is Enif which is said to mark the horse's nose.

PISCES, The Fish

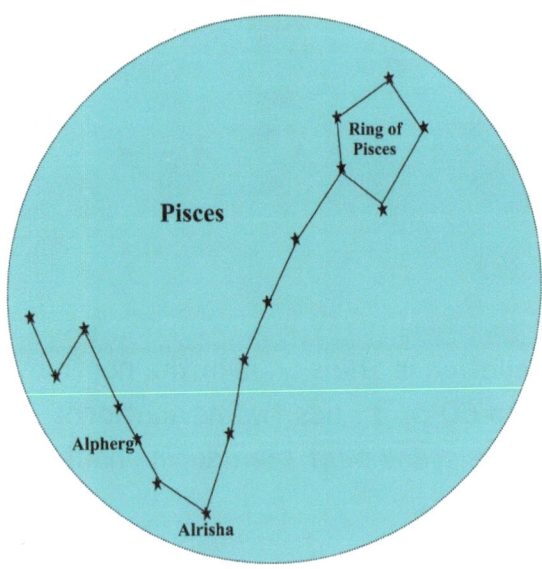

Pisces is a large 'V' shaped constellation which straddles the equator and lies on the path of the ecliptic. It is visible between latitudes 90°N and 65°S; it is best seen in November. The brightest star in Pisces is Alpherg or Kullat Nunu but this is not a navigational star; in fact, this constellation contains no navigational stars. The reason that we have included Pisces in this 'route map' is because of its association with the 'First Point of Aries' which is the point at which the Sun crosses the celestial equator when it is moving from south to north along the ecliptic. This event occurs on 21/22 March and is known as the vernal Equinox. The confusing thing is that, although the 'First Point of Aries' lay in the constellation of Aries when it was chosen by the ancient astronomers, due to precession it now lies in Pisces.

The name Pisces is derived from the Latin for fish and is said to depict two fish, swimming in opposite directions, held together by a piece of string connecting their tails. The star Alrisha is said to be the knot that ties the string that hold the two fish together. In ancient Greek mythology, Pisces is associated with the fish that carried Aphrodite and Eros to safety from the monster Typhon. In another mythological tale, the fish of Pisces were said to have been spawned by the 'Great Fish' in the constellation Pisces Austrinus which is known as the 'Southern Fish'.

Finding Pisces.

Pisces lies just to the south of the 'Great Square of Pegasus as shown in the diagram which follows. If a line is drawn from Scheat to Algenib in Pegasus and extended by about one hand-span, it will point to the star Alrisha in Pisces; however, this constellation is very hard to find because it is so faint.

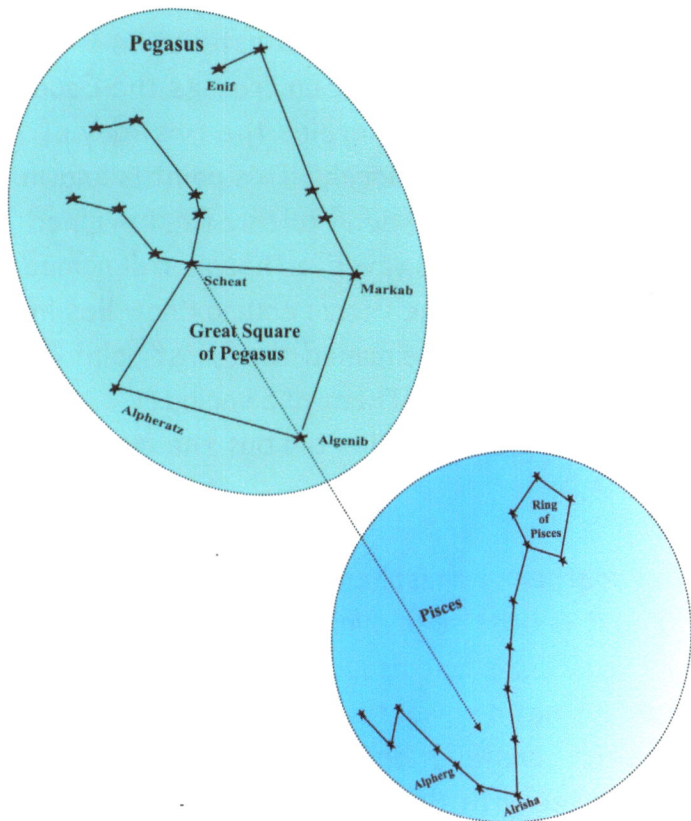

Aquarius, the Water-Bearer

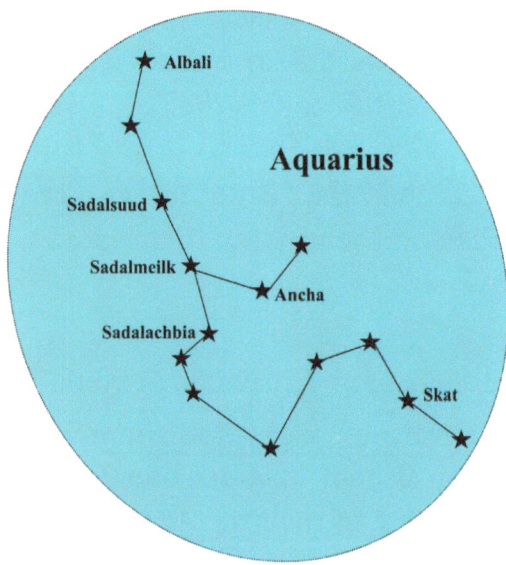

In a popular song, the words the 'the dawning of the age of Aquarius' refer to the period when the vernal equinox will lie inside the constellation Aquarius. The vernal equinox is the point where the Sun crosses the Equator on its northward movement along the ecliptic and heralds the first day of spring in the northern hemisphere on 20th./21st. March. This point is known as the 'First Point of Aries' because in 150 B.C. when Ptolemy first mapped the constellations, Aries lay in that position. However, although still named the 'first point of Aries', due to precession, the vernal equinox now lies in the constellation Pisces, so logically, it should be named the 'first point of Pisces' since we are now in the 'Age of Pisces'. There are various predictions of when the next 'age of Aquarius' will begin but the most prominent of these is about 2600 A.D.

Aquarius is a constellation in the southern hemisphere and is visible at latitudes between 65°N and 90°S; it is best seen during the month of October. The brightest star in Aquarius is Sadalsuud an Arabic phrase meaning "luck of lucks". Sadalsuud is not a navigational star and in fact, there are no navigational stars in the constellation Aquarius which, like Pisces, is very faint and difficult to see with the naked eye.

In ancient Greek mythology, Zeus transformed himself into an eagle (Aquila) to carry a young man named Ganymede to serve as a cup-bearer to the gods in Olympus. The name Aquarius is derived from the Latin for 'water-bearer' or 'cup-bearer'.

Finding Aquarius.

Aquarius is a very faint constellation and is difficult to locate. However, this diagram shows that if we line up the two stars that form the base of the triangle at the top of the 'Ring of Pisces' and extend that line it will point to the star Sadalmeilk in the constellation Aquarius which is to the south of Pegasus.

If we also run a line from Scheat to Markab in Pegasus and extend that line by a palm-width, that too will point to Aquarius.

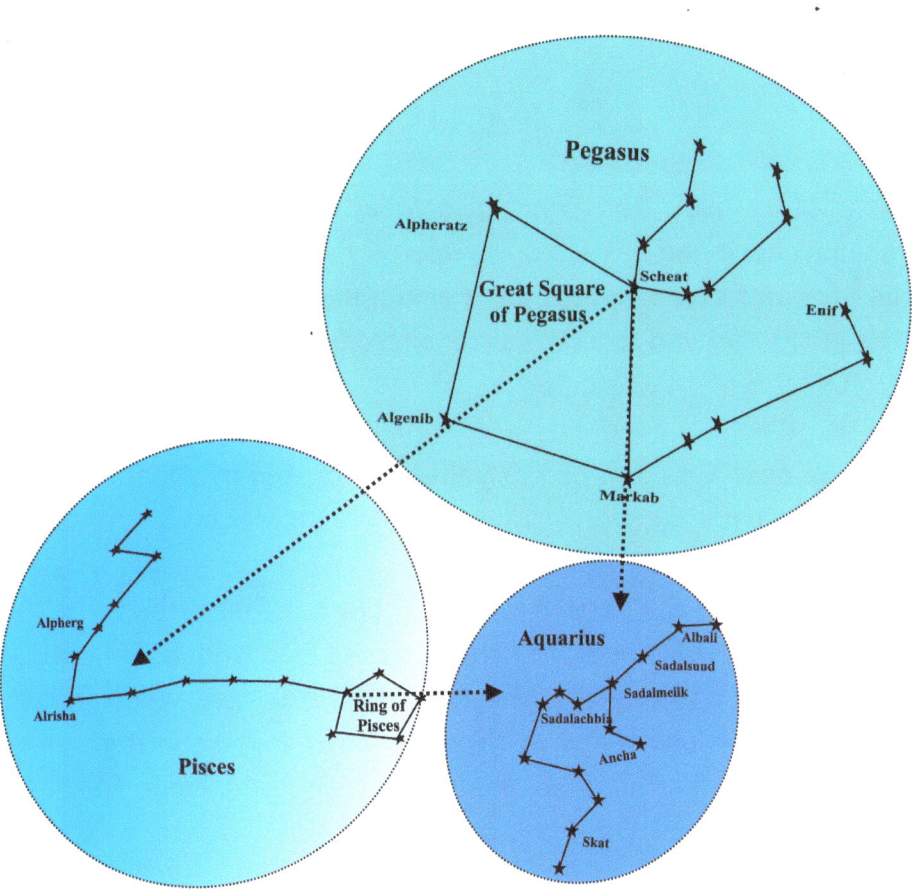

Piscis Austrinus, the Great Fish or the Southern Fish

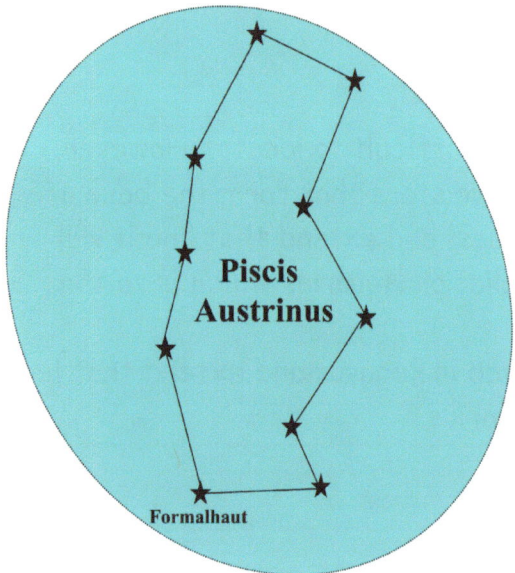

The small constellation Piscis Austrinus, also called Piscis Australis, lies in the southern hemisphere and is visible between latitudes 55ºN and 90ºS. It contains mostly faint stars except for **Formalhaut** which is one of the brightest stars in the sky and is a navigational star. Because it is the only bright star in this part of the sky, Formalhaut is sometimes called the "Lonely Star of Autumn".

For navigators, the best time to see Formalhaut is during nautical twilight is in the month of October.

The ancient Greeks named the constellation the 'Great Fish' which, according to Egyptian mythology, saved the goddess Isis who, as a reward, sent it into the sky where it spawned the two fish in the constellation Pisces. Piscis Austrinus is also associated with the Babylonian myth about the goddess Atargatis who fell into a lake and was rescued by a large fish. The name Piscis Austrinus is derived from the Latin for the 'Southern Fish'.

Finding Piscis Austrinus

Fomalhaut was once considered to be part of the constellation Aquarius as well as the constellation Pisces Austrinus, where it now belongs. Formulhaut is depicted as the toe of Aquarius and this idea provides us with a way of locating both Aquarius and Pisces Austrinus for if we can find Formalhaut, the brightest star in the region, then we can find both of those constellations.

The following diagram shows Pisces Austrinus nestling at the foot of Aquarius with Formalhaut providing the link between them.

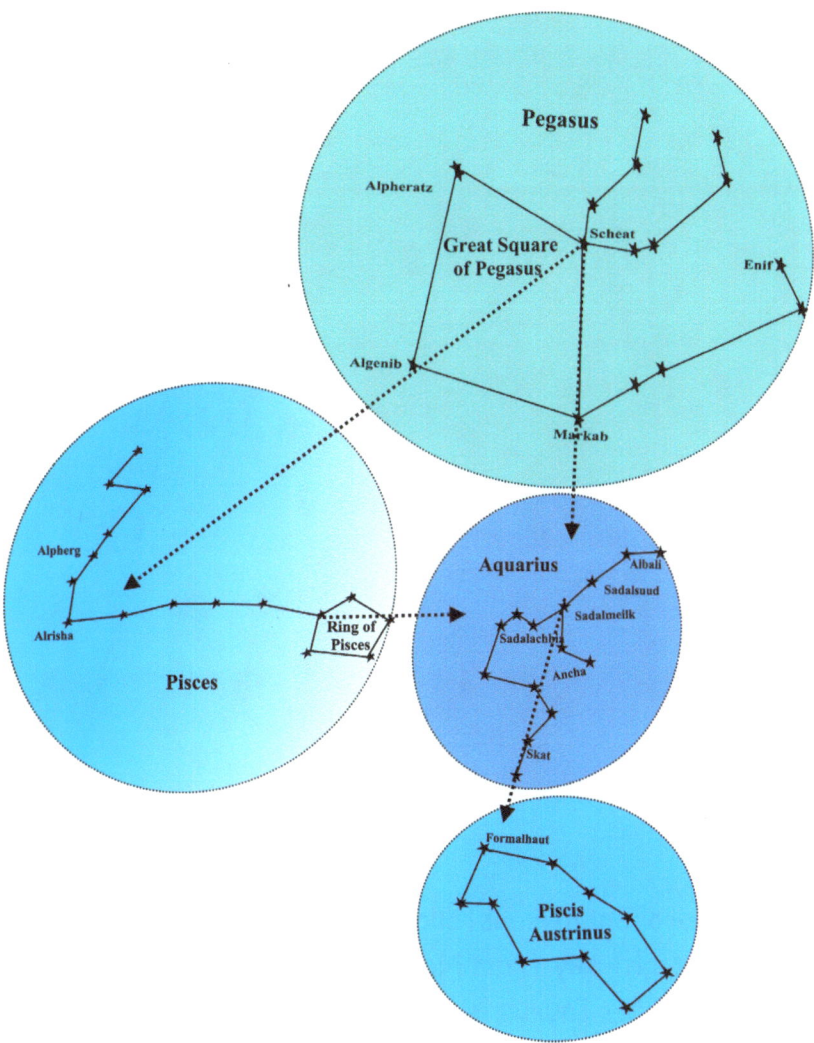

Winter Stars in the Northern Hemisphere (Summer Stars in the Southern Hemisphere).

The Autumn constellations, Pegasus, Andromeda, Pisces, Aquarius and Piscis Austrinus sink into the west as we move into winter and new constellations take their place in the night sky of the northern hemisphere including Taurus, Orion, Canis Major and Minor, Gemini and Aries.

Taurus, The Bull.

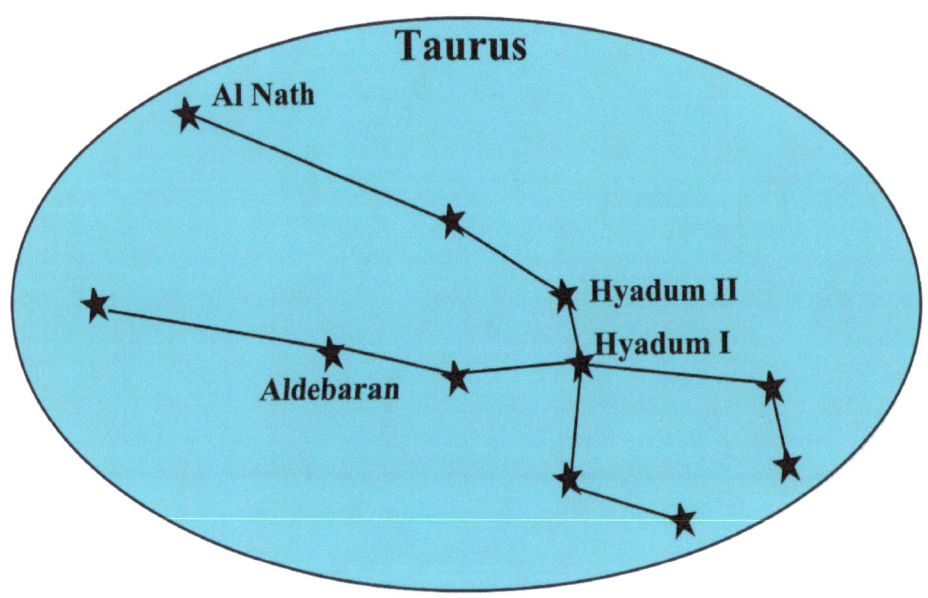

Taurus is one of the most prominent constellations in the northern winter sky. It passes through the night sky from November to March and is visible at latitudes from 90°N to 65°S. Taurus is most visible in January and for the benefit of navigators, it is always on hand for star sights during nautical twilight throughout that month.

Taurus is most famous for its red giant star, **Aldebaran,** which is known as Taurus' Eye, is the 14th brightest star in the sky and is an important navigational star.   The star at the tip of the northern horn of Taurus, **Al Nath** (sometimes spelled El Nath) is the second brightest star in the constellation and is also a navigational star. In earlier times this star was considered to be shared with the constellation Auriga, forming the right foot of the Charioteer as well as the Northern horn of the bull.

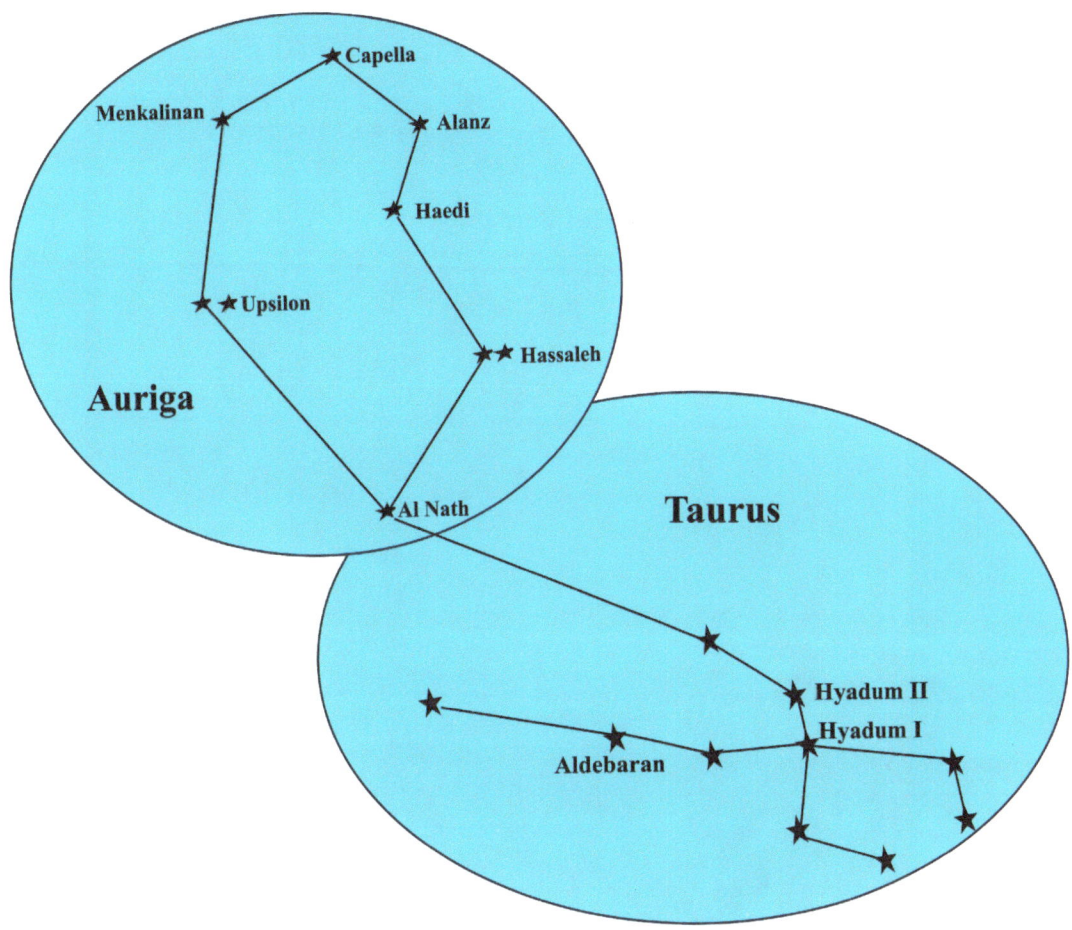

Taurus is associated with several mythological beliefs. In Greek mythology, Zeus was said to have disguised himself as a bull to abduct Europa, the daughter of king Agenor. The ancient Egyptians believed that the constellation represented the sacred bull associated with spring and Babylonian astronomers called it the 'Heavenly Bull'.

Finding Taurus.
If we imagine a line from Phad to Meral in Ursa Major and extend it for a distance of 80° or roughly 4 hand-spans it will point directly to the star

Aldebaran in the constellation Taurus. Once we have located Aldebaran the remaining stars of Taurus can easily be identified.

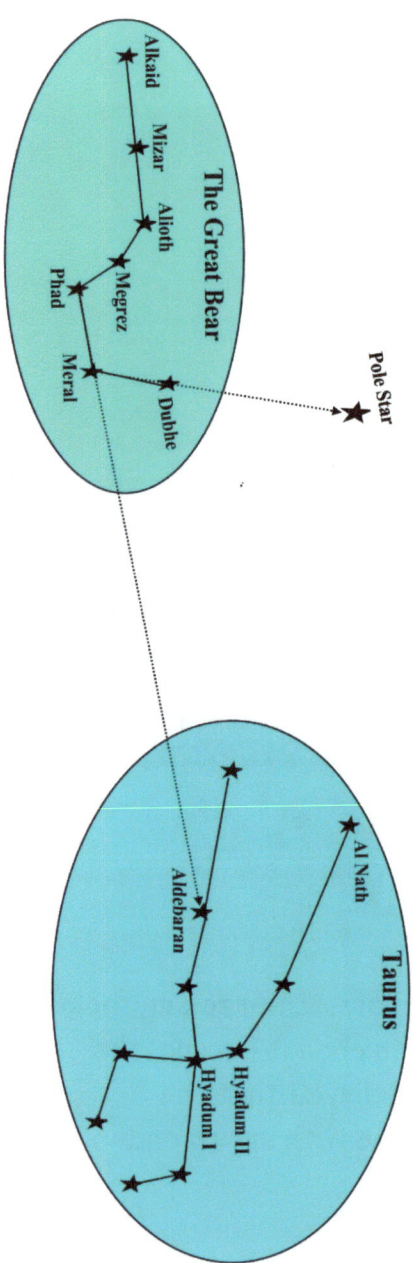

Auriga the Charioteer

The constellation Auriga is in the northern hemisphere and is visible between latitudes 90°N to 40°S. For navigator's, it is best seen during nautical twilight in February and March.

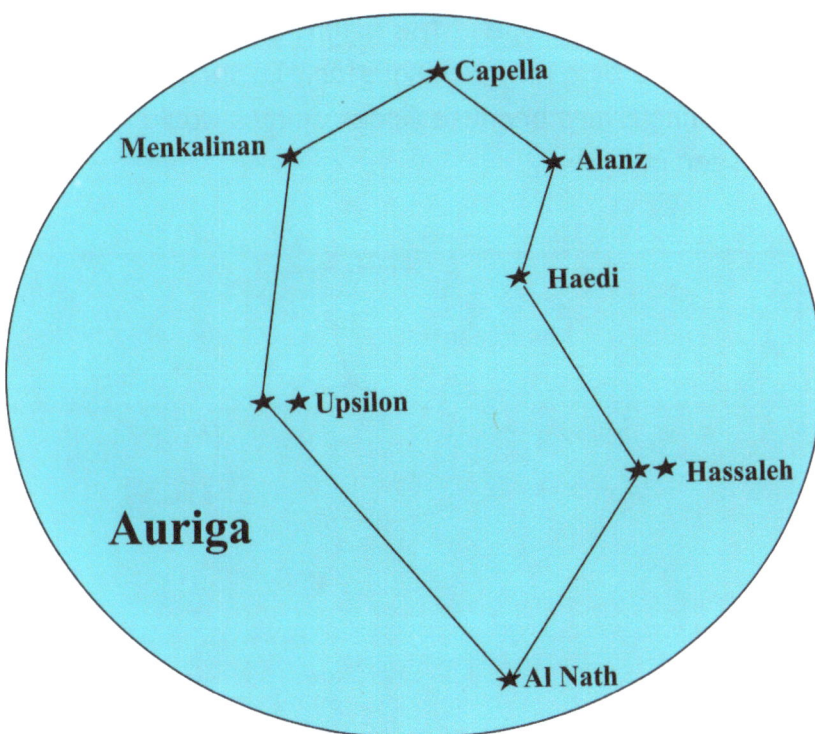

Capella is one of the navigational stars; it is the brightest star in the constellation Auriga and the 6th brightest in the northern hemisphere.

The star Al Nath forms the right foot of the Charioteer and as has already been explained, it was once considered to be shared with the constellation Taurus where it forms the northern horn of the bull. If we still think of Al Nath (the second brightest star in Taurus and also the second brightest in Auriga) as the foot of Auriga, we will have an easy method of finding both Auriga and Taurus.

In mythology, Auriga is associated with Myrtilus the charioteer because the shape of the constellation was said to resemble a pointed helmet of a charioteer. It is also identified with Hephaestus, the god of the blacksmiths who invented the chariot.

The Pleiades, the 'Sailing Stars' or the 'Seven Sisters'.

If we draw a line from Aldebaran to Tau Tauri in Taurus (the two eyes of the Bull) this will point roughly in the direction of the Merope in the star cluster Pleiades which lies about 10° (a palm-width) from Taurus.

The Pleiades cluster is visible between latitudes 90°N and 65°S and is best seen during the month of January. Merope is the brightest of these stars but even so, it not considered to be a navigational star. In ancient times, the Pleiades were called the Sailing Stars because Greek sailors would not put to sea unless they could be seen in the sky.

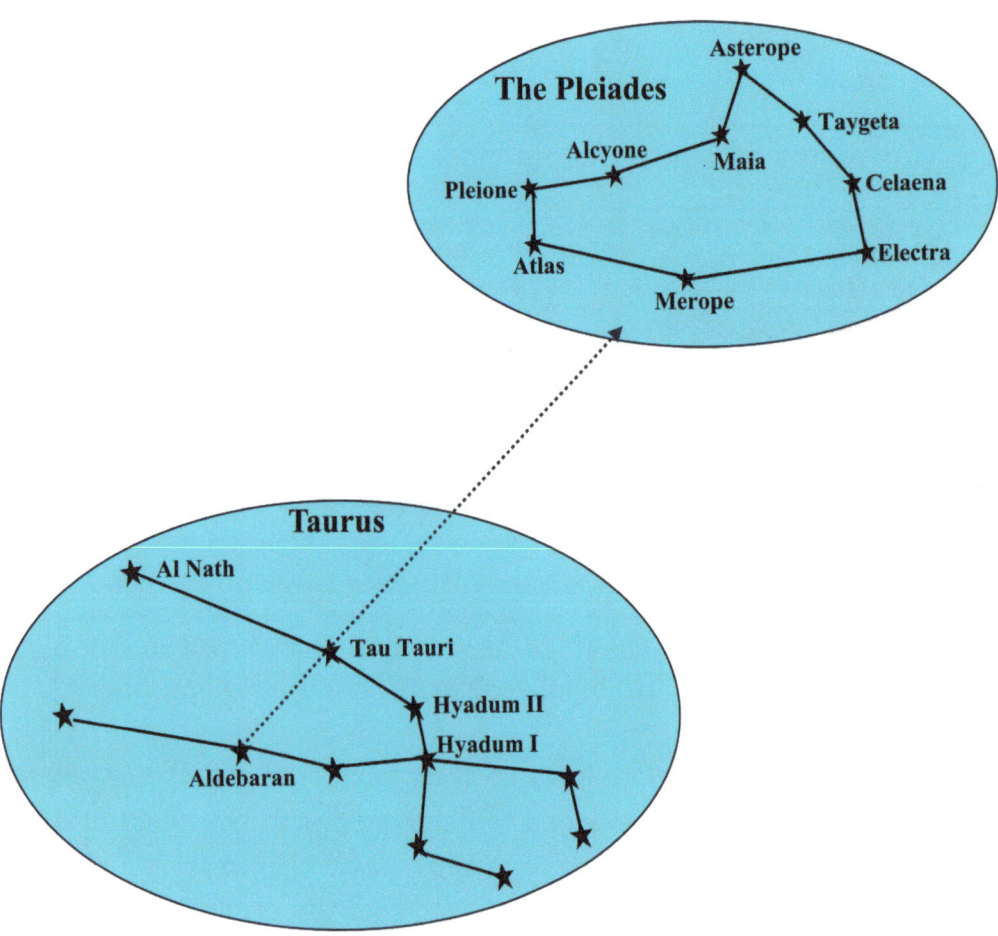

Another name for the Pleiades is the 'Seven Sisters' after Alcyone, Maia, Anterope, Taygeta, Celaena, Electra and Merope from Greek Mythology. Although Pleione was not one of the seven sisters, she along with her consort Atlas is included in the Pleiades Cluster.

The Winter Triangle (Orion, Canis Major and Canis Minor).

Orion, The Hunter

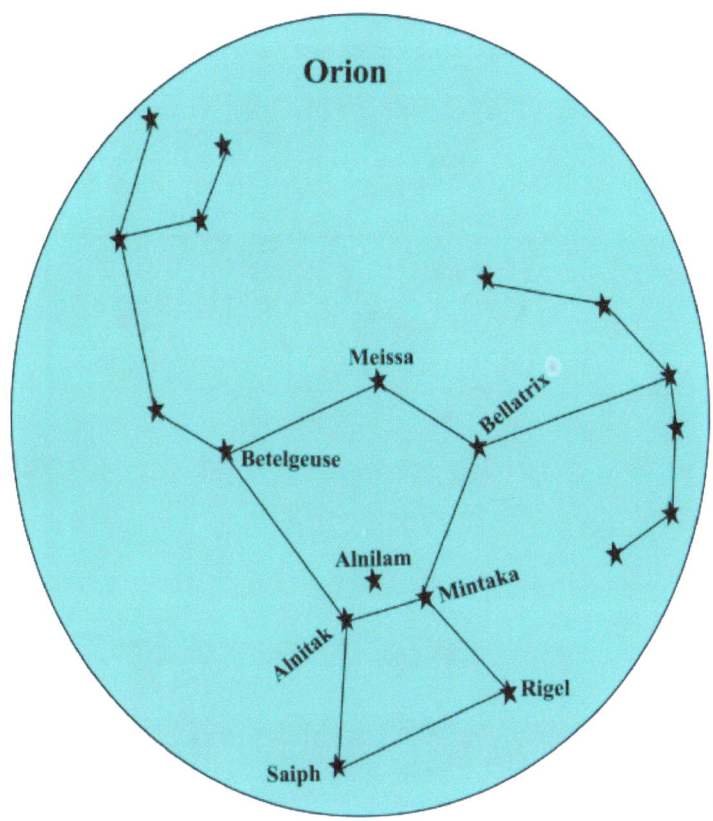

Orion is one of the brightest and best known constellations in the night sky it straddles the equator and is visible between latitudes 95°N and 75°S. This easily recognized constellation contains 4 navigational stars, **Rigel, Belatrix, Anilam** and **Betelgeuse** which, for navigation purposes are best seen during nautical twilight in the month of January.

In Greek mythology, this constellaion represents the mythical hunter Orion, who is often depicted in star maps as facing the charge of Taurus, the bull, pursuing the Pleiades sisters with his two hunting dogs which are represented by the nearby constellations Canis Major and Canis Minor. Meissa marks the position of the Hunter's head while Betelgeuse, and Belatrix are his shoulders. Alnitak, Alnilam and Mintaka form his belt and from this hangs his sword which is marked by the Orion Nebula. His right thigh is marked by Saiph and Rigel marks his left foot.

Finding Orion

If we take a line from Tau Tauri to Aldebaran in the constellation Taurus and extend this line for roughly three hand-spans we will come to the star Belatrix in Orion as the diagram below shows.

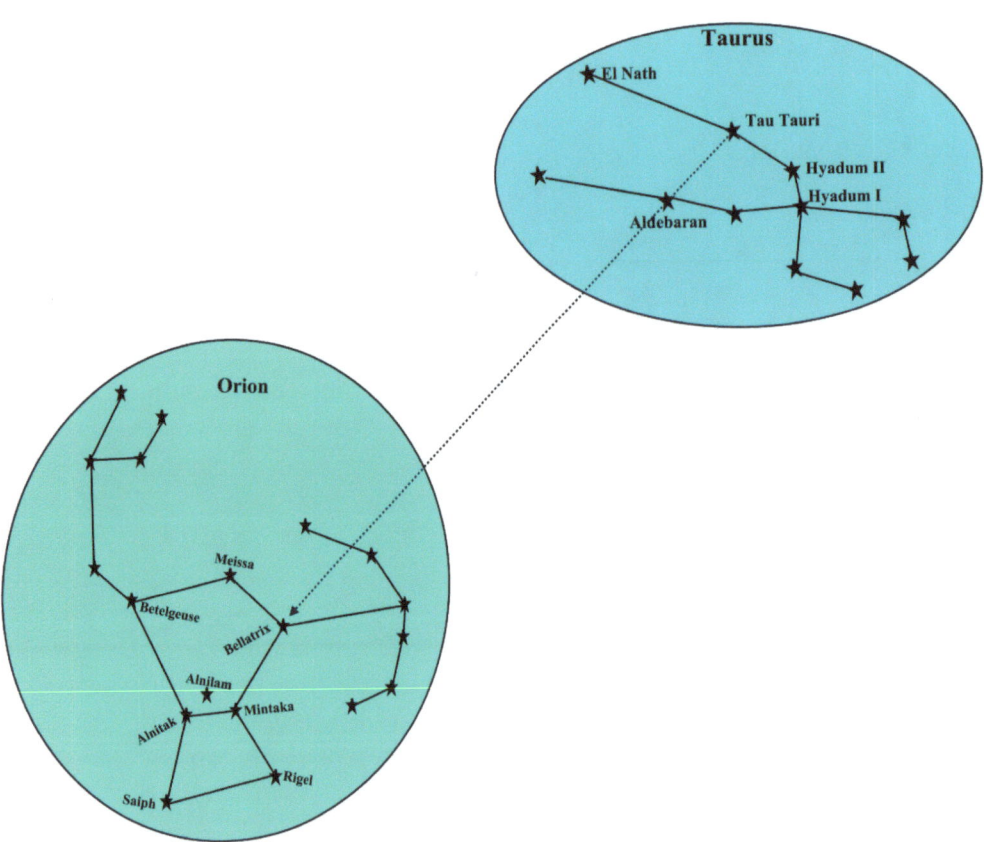

Canis Major, The Greater Dog

The constellation Canis Major is in the Southern Hemisphere; it is visible from 90°S to 60°N and is best visible from November to March. As shown in the diagram below, Canis Major contains Sirius, which is otherwise known as the Dog Star. **Sirius** is the brightest star in the sky and is a navigational star which for navigation purposes is best seen during nautical twilight during the month of February.

Canis Major is Latin for 'The Greater Dog' and is so named because it was said to represent one of the two hunting dogs of Orion the Hunter; the other dog being Canis Minor. In Greek mythology, Zeus sent Laelaps, an alternative name for Canis Major, into the sky when it failed to outrun a fox.

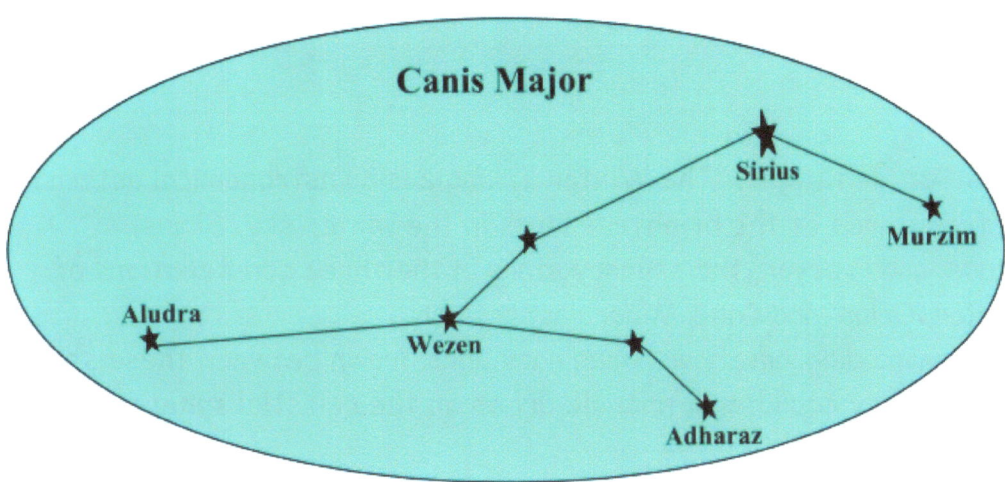

Canis Minor The Lesser Dog

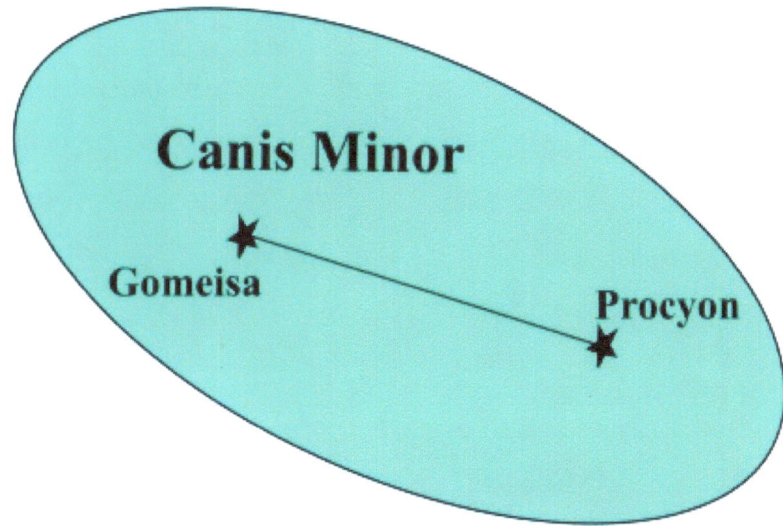

Canis Minor is a small constellation in the northern hemisphere and is visible at latitudes from 85°N to 75°S.

Its name means 'the lesser dog' in Latin and it is said to represent one of the dogs that follow Orion. In another mythological tale, Canis Minor represents Maera the dog which was sent to the sky by Zeus after it died of grief when its master Icarius was killed.

There are only two bright stars in Canis Minor, Gomeisa and **Procyon** which is a navigational star and which, for navigational purposes is best seen during nautical twilight in March.

The Winter Triangle. The Winter Triangle is an astronomical asterism formed from three of the brightest stars in the winter sky; Sirius, Betelgeuse, and Procyon, the primary stars in the three constellations of Canis Major, Orion, and Canis Minor respectively.
As the following diagram shows, imaginary lines drawn between these stars form an imaginary equilateral triangle drawn on the celestial sphere,

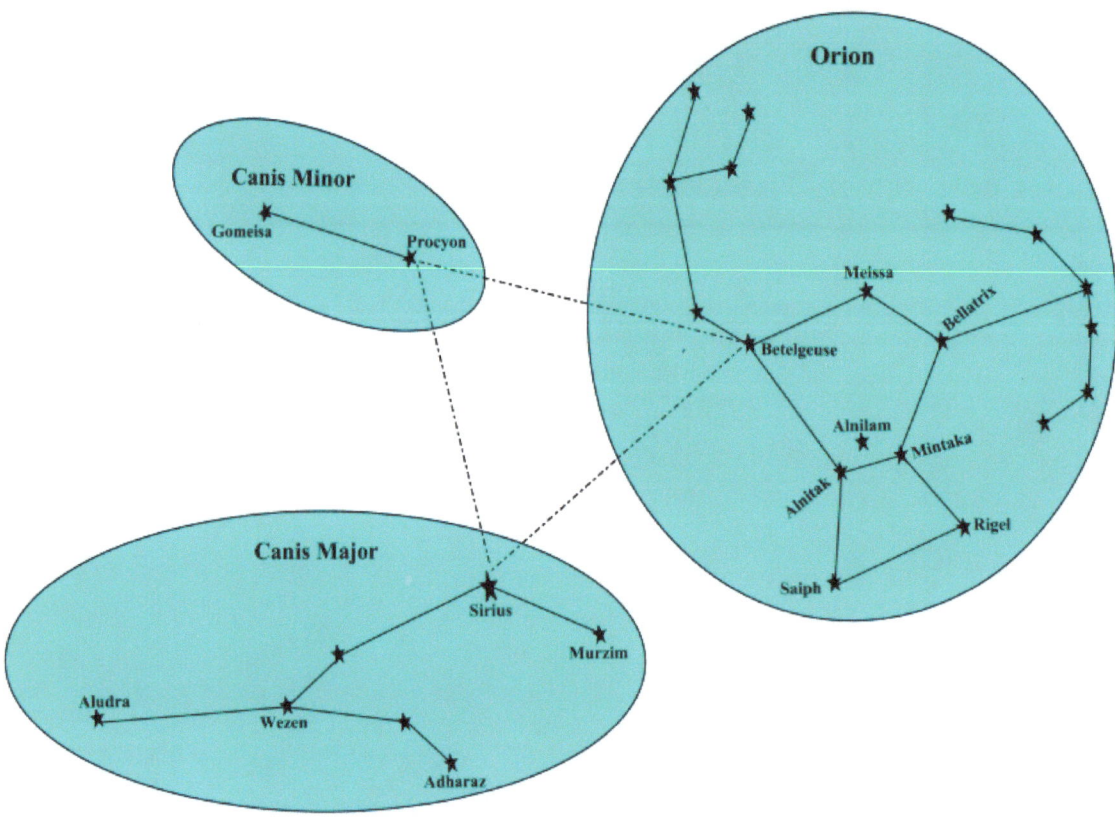

Gemini, The Heavenly Twins.

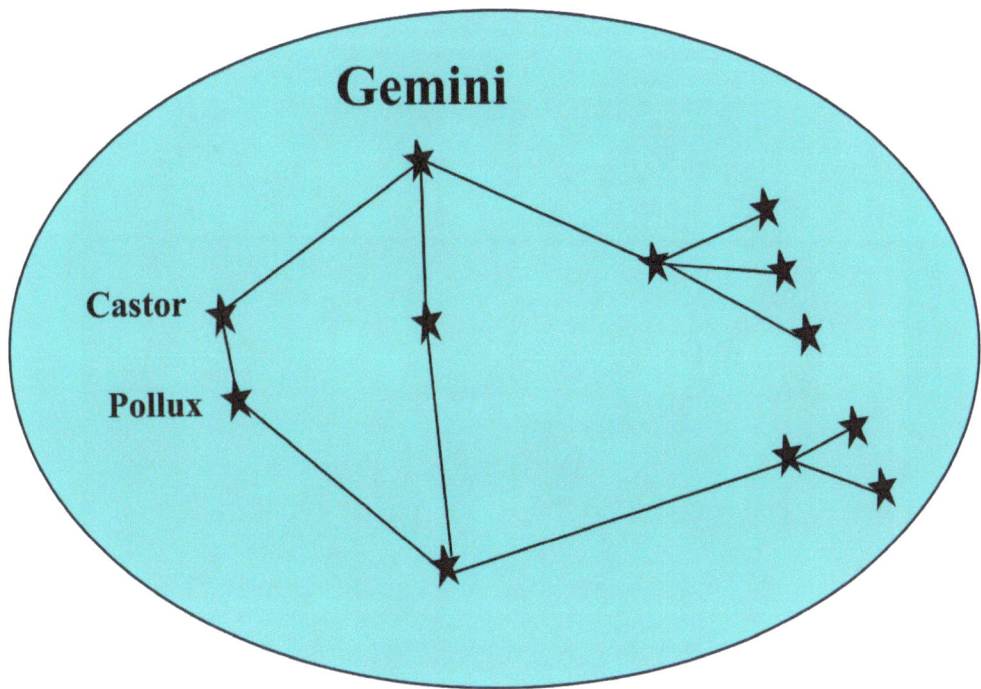

The name Gemini means Twins in Latin and for this reason, it has the alternative name the 'Heavenly Twins'. The constellation is associated in, Greek mythology, with the twins Castor and Polydeuces. Pollux and Castor, the brightest stars in the constellation are said to form the eyes of the twins. (We use the Latin name Pollux instead of the Greek name Polydeuces).

Gemini is in the northern hemisphere and is visible between 90°N and 60°S. **Pollux** is a navigational star and is best seen for navigation purposes during nautical twilight in February.

Finding Gemini. As shown below, a line running from Megrez to Merak in Ursa Major leads to the constellation Gemini.

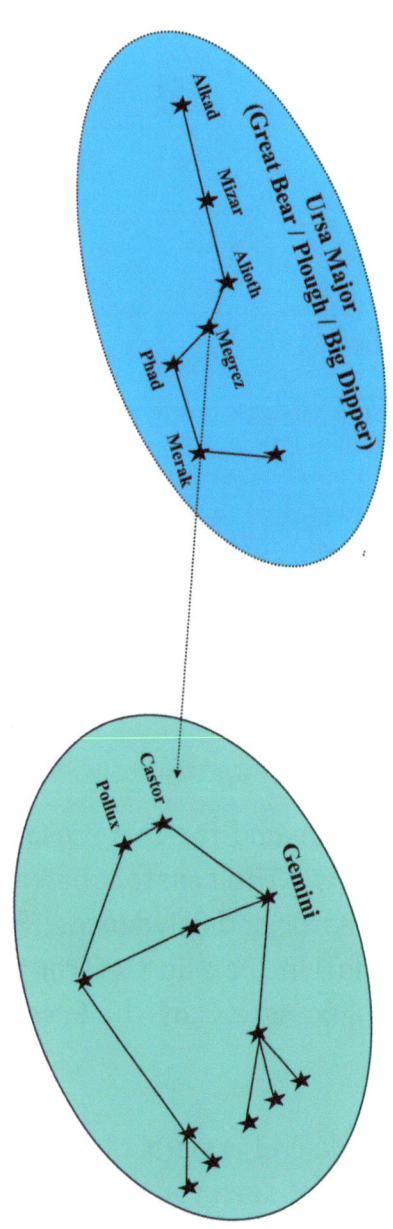

Aries, The Ram

The constellation Aries is located in the Northern Hemisphere and is visible between latitudes 90°N and 60°S. The name Aries, which means Ram in Latin, is associated with Jason and the Golden Fleece in Greek mythology.

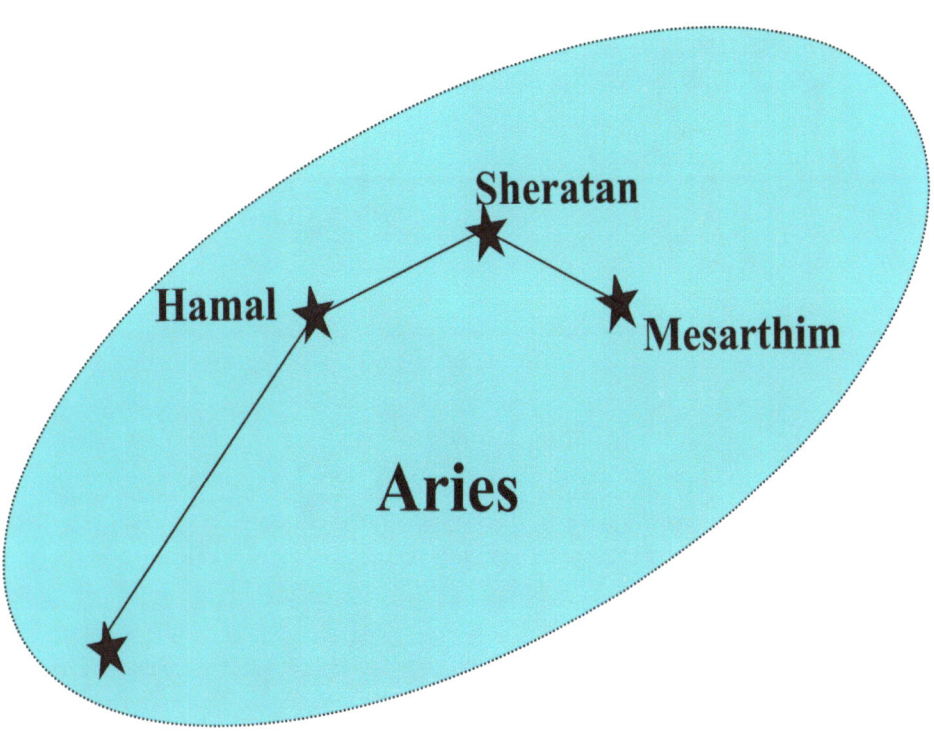

Aries is a difficult constellation to see with the naked eye but it can be found about mid way between the Pleiades and the constellation Pegasus. Pegasus lies to its west, Pleiades to its east as the following diagram shows. (Remember that, in star maps, east and west are reversed).

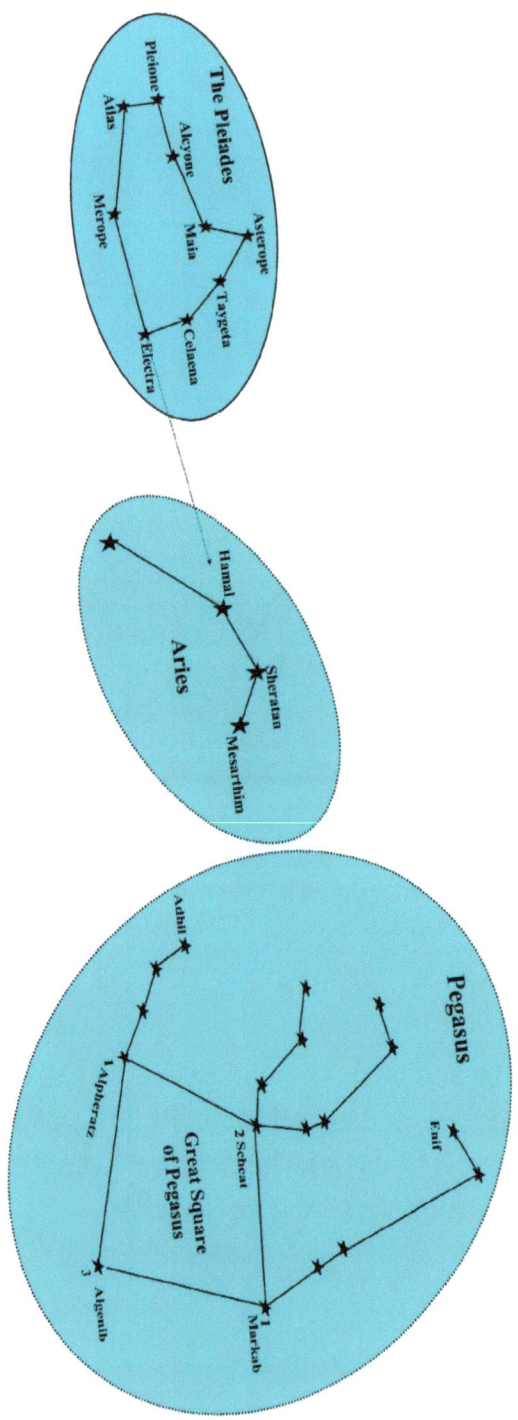

The First Point of Aries. As discussed in chapter 2, just as the Greenwich meridian has been arbitrarily chosen as the zero point for measuring longitude on the surface of the Earth, 'the first point of Aries' was chosen as the zero point in the celestial sphere. It is the point at which the Sun crosses the celestial equator moving from south to north (at the vernal Equinox in other words). The confusing thing is that, although this point lay in the constellation of Aries when, in 150 B.C., Ptolemy first mapped the constellations, due to precession, it now lies in Pisces. The First Point of Aries is usually represented by the 'ram's horn' symbol shown below:

Hamal is the brightest star in Aries and it is a navigational star which, for navigation purposes, is best seen during nautical twilight in the month of December.

Spring Stars in the Northern Hemisphere (Autumn Stars in the Southern Hemisphere).

Spring is just around the corner and when the winter constellations begin to make their exits to the west, we will find Boötes, Leo, Cancer, Hydra and Virgo lurking in the wings ready to take their places. They will soon join the ever-present circumpolar constellations of Ursa Major and Minor, Cassiopeia and Auriga in the northern hemisphere's night sky.

Leo, The Lion.

Leo is one of the largest constellations in the sky and is visible throughout the northern hemisphere and the northern regions of the southern hemisphere during the spring and early summer months. Leo is home to two navigational stars, **Denobola** and **Regulus** which are shown in the diagram below. Regulus, the brightest star in the constellation, is said to mark the lion's heart with Denobola marking the tip of its tail. From a navigation perspective, these stars are best seen during nautical twilight during the month of April.

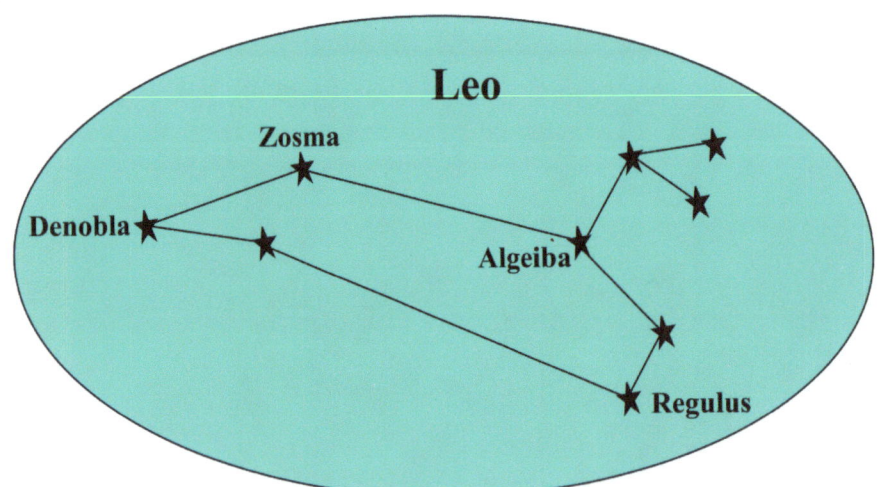

The name Leo means Lion in Latin and the constellation, which is depicted as a crouching lion, is associated in Greek mythology with the lion of Nemea which was killed by Heracles as one of his twelve labours.

How to find Leo.   When the line of pointers in Ursa Major is produced in the opposite direction to the Pole Star, that is from Dubhe to Merak, it will point to Leo as shown in the diagram below.

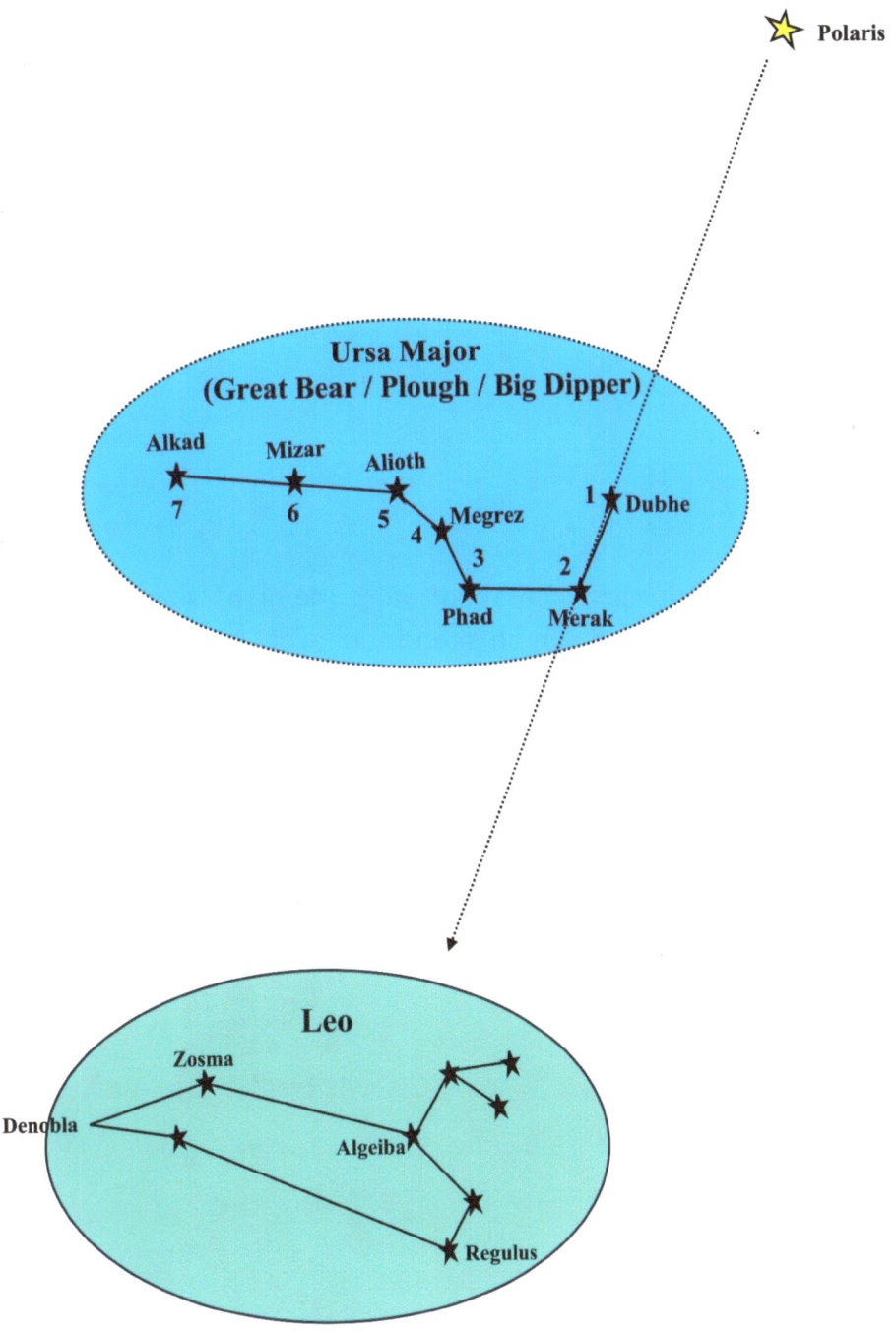

Boötes The Herdsman.   If we take a line from Alioth to Alkaid in the Great Bear and extend that line in an imaginary curve for about roughly three hand-spans as shown in the diagram below, it will point to Arcturus, the brightest star in the constellation Boötes. Arcturus is the fourth brightest star in the sky and a navigational star. The ancient Greeks named Arcturus the "Bear Watcher" because it seems to be looking at the Great Bear (Ursa Major).

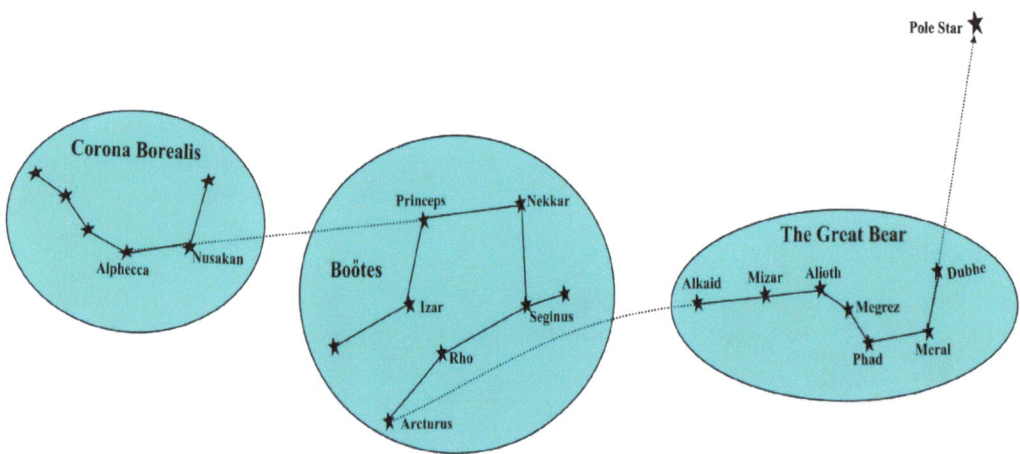

Boötes is the 13th largest constellation in the night sky; it is located in the northern hemisphere and can be seen from 90°N to 50°S. The ancient Greeks visualized it as a herdsman chasing Ursa Major round the North Pole and its name is derived from the Greek for "Herdsman".

Corona Borealis, The Northern Crown. If we next take a line from Nekkar to Princepes in Boötes, and extend that line by about one and a half hand-spans, it will point to Nusakan in the close by Corona Borealis constellation.

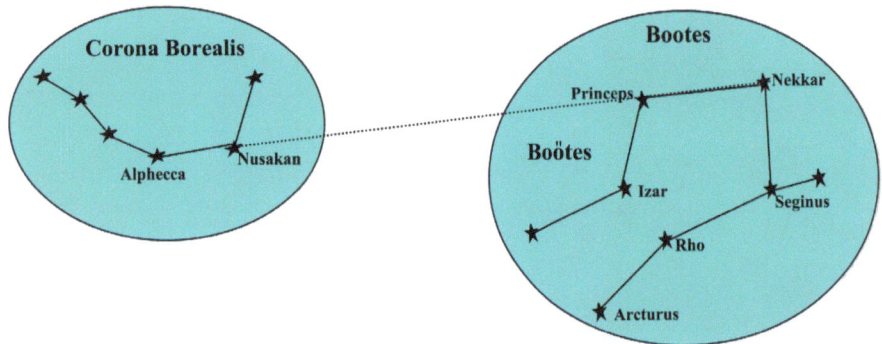

Corona Borealis, whose name in Latin means 'northern crown', is a small constellation in the northern hemisphere and can be seen between latitudes 90°N and 50°S.

The main stars in Corona Borealis form a semi-circle which is associated with the crown of Ariadne in Greek mythology. It is said that the crown was given by Dionysus to Ariadne on their wedding day and after the wedding, he threw it into the sky where the jewels became stars which were formed into a constellation in the shape of a crown. Alphecca, the brightest star in the group, is a navigational star and can be seen during nautical twilight from April to July.

Virgo, the Virgin. As we learned when studying the constellation Boötes, an imaginary curved line from Alkaid in the Great Bear leads to the constellation Boötes. If, as shown in the following diagram, we continue that curved line by another hand span from Arcturus we will come to the bright bluish-white star Spica in the constellation Virgo. Virgo lies over the southern hemisphere and is one of the largest constellations in the sky; it is visible between latitudes 80°N and 80°S.

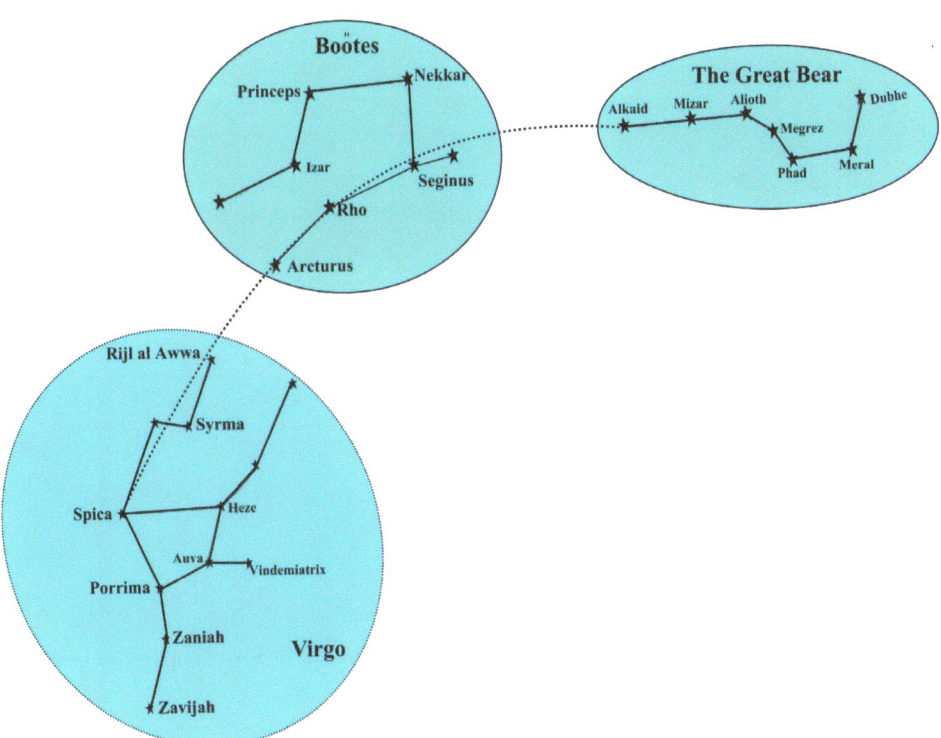

The brightest star in Virgo is **Spica**, the 15th brightest star in the sky and a very important navigational star which can be seen during morning nautical twilight from December through to May and during evening nautical twilight

Page 99

from April through to September.

In ancient Greek mythology, Virgo is associated with the goddess Dike, the goddess of justice and the constellation Virgo takes its name from the Latin for virgin or young maiden.

Libra, The Scales. This is one of the hardest constellations to find because it has no bright stars. The ancient Greeks considered it to be part of the Scorpio constellation because it was said to represent the claws of the scorpion and this gives us the clue to locating Libra in the sky. The diagram below shows the close proximity of Libra to Scorpius and it is easy to see how the representation described above was imagined.

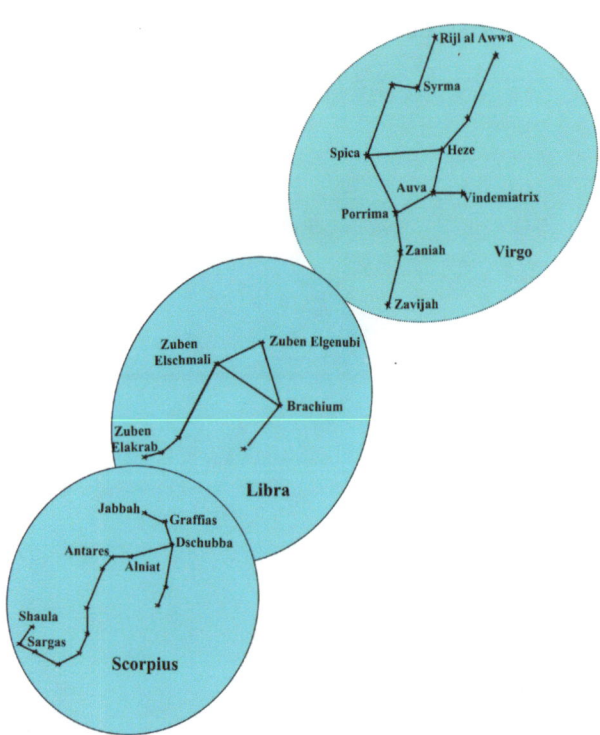

Libra lies in the southern hemisphere between 10° and 30° south and can be seen between latitudes 65°N and 90°S.

The name Libra means 'The Weighing Scales' in Latin and the constellation Libra is depicted as the scales of justice held by the Greek goddess of justice, Dike who is represented by the constellation Virgo. In the diagram below, the stars Zuben Elschmali and Zuben Elgenubi mark the scales' balance beam with Elakrab and Brachium representing the weighing pans.

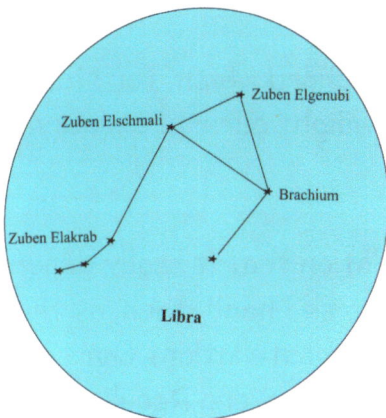

As a matter of interest, the star name Zuben Elgenubi may have been the inspiration for the name 'Obi-Wan Kenobi' of Star Wars fame because of their similar sounds.

Zuben Elschmali is the brightest star in the constellation but it is not considered to be a navigational star. In fact there are no navigational stars in Libra.

Hydra, the sea serpent.

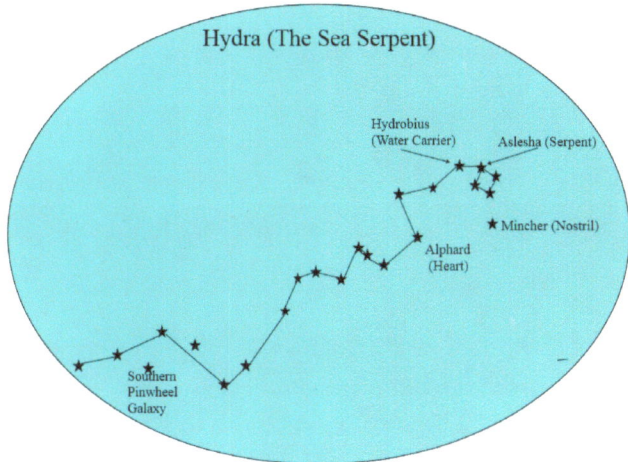

The constellation Hydra, the sea serpent, is the largest constellation in the night sky and it is also one of the longest constellations. Hydra's head is located south of the constellation Cancer and its tail lies between Centaurus and Libra. It is best seen from the southern hemisphere, but can be observed in the northern hemisphere between January and May.

Hydra contains one navigational star and that is Alphard which, for star sights, is best observed during evening nautical twilight during February, March and May.

Finding Hydra. Hydra is such a large constellation that it really depends on which part of it you want to see. If you want to see head, then, as the diagram below shows, you should look between the constellations Canis Minor and Leo (look for the bright stars Procyon in Canis Minor and Regulus in Leo).

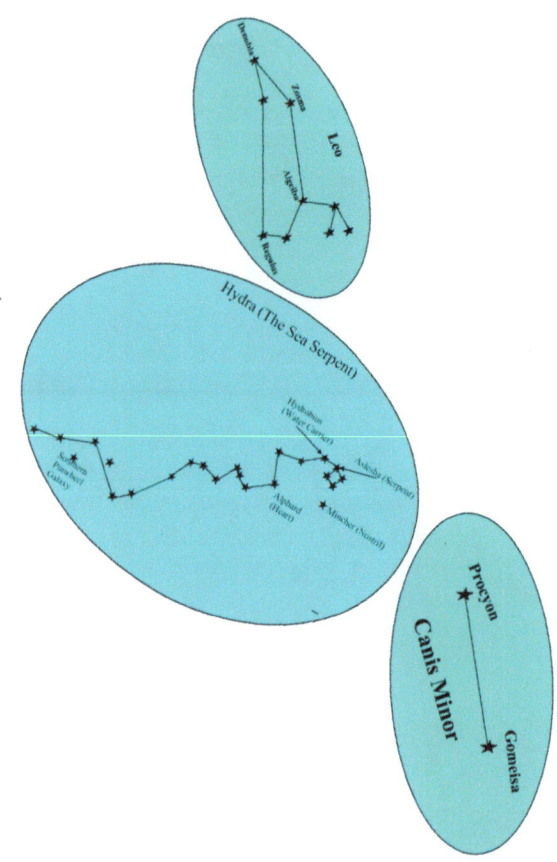

In Greek mythology, Hydra represents the water snake brought to the god Apollo by the crow Corvus as an excuse for being late from his errand to fetch water. It may also represent the hydra from the myth of Hercules and his twelve labours. The Hydra was a giant beast with the body of a dog and 100 snake-like heads. It was slain by Hercules on the second of his twelve labours for the king of Mycenae. As each head was cut off, two more would grow in its place. Hercules burned the roots of the heads to prevent them from growing back.

Cancer The Crab.   Cancer is a relatively small constellation in the northern hemisphere and is visible between latitudes 90°N and 60°S.

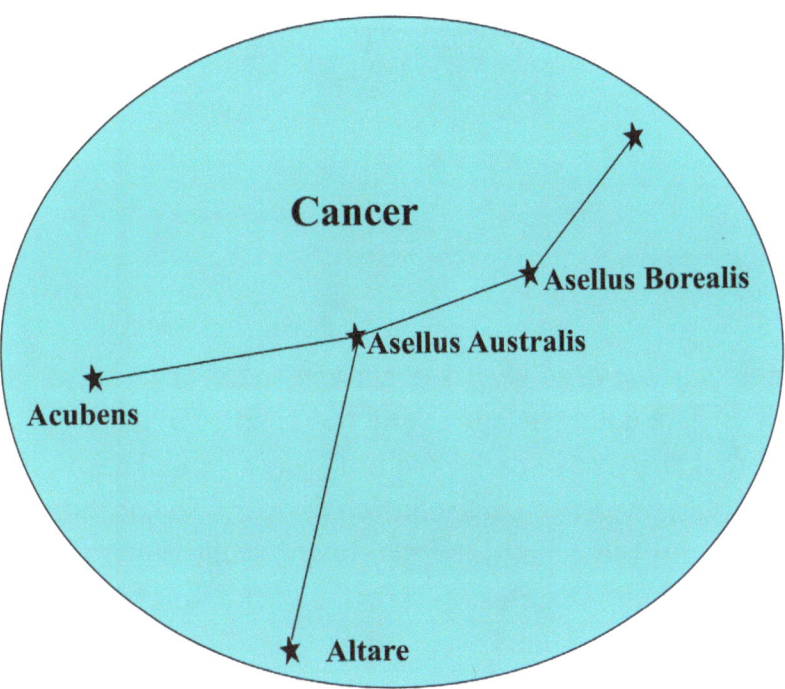

Cancer consists of mainly faint stars, none of which is a navigational star and for this reason, it is not a very useful constellation for astro navigation. However, it does help us in one way: Although astrology has no place in astro navigation, the signs of the zodiac can be very useful to navigators because the order in which they follow one another can tell us the position of one zodiac constellation in the sky with respect to another. For example, we know that Cancer's position on the ecliptic falls between Leo and Gemini and as the following diagram demonstrates, Cancer can easily be found nestling between those constellations with Gemini lying to the west and Leo to the east (remember that in star maps, east and west are reversed with respect to conventional maps).

From this, it can be seen that Cancer can be an aid to locating both Gemini and Leo.

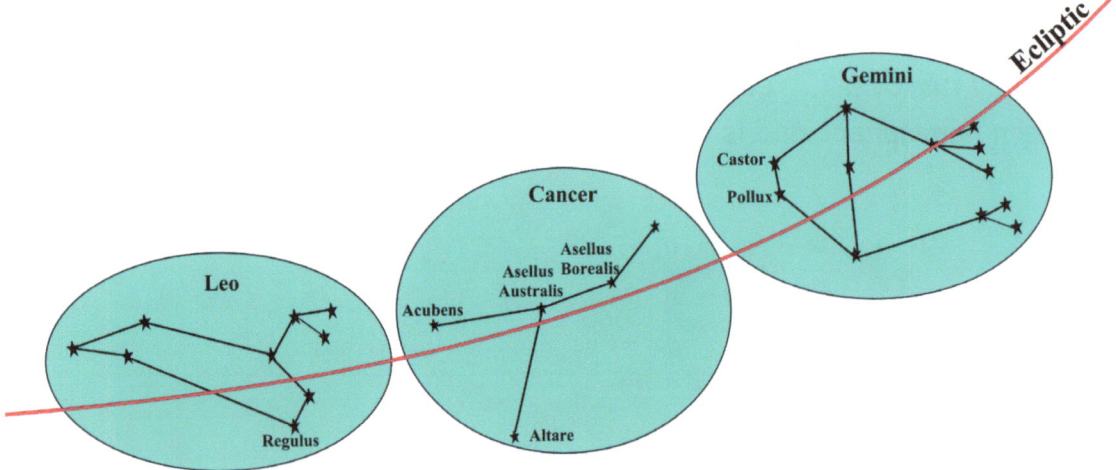

In Greek mythology, Cancer is associated with the crab in the story of the Twelve Labours of Heracles. The goddess Hera sent the crab to attack Heracles while he was fighting the Lernaean Hydra but Heracles kicked it all the way to the stars where it formed the constellation Cancer. In another version of the story, Hera placed the crab in the sky in gratitude for its efforts even though it was killed by Heracles. (Heracles is the Roman name for the Greek god Hercules).

The Tropic of Cancer. These days, the Sun passes through Cancer in late July; however, in the time of Ptolemy, around 2000 years ago, this occurred during the summer solstice when the Sun reached 23.4° N, the northern limit of the ecliptic. The latitude 23.4° N is still called the Tropic of Cancer even though the Sun now resides in Taurus at the summer solstice.

Perseus. Although the constellation Perseus which features on page 54 is considered to be circumpolar in northern parts of the northern hemisphere, it becomes visible throughout the whole of the northern hemisphere and northern parts of the southern hemisphere during spring and early summer.

Chapter 6
Latitude and Longitude

A good starting point for a study of latitude and longitude is to consider the definitions of Great Circles and Small Circles:

Great Circles. A plane of a sphere which passes through the centre of the sphere is called a great circle. The paths of two great circles of a sphere will intersect at two points 180° apart on the surface of the sphere.

Small Circles. A plane of a sphere which does not pass through the centre of the sphere is called a small circle. Two small circles of a sphere need not meet.

Now that we have established these definitions, we can begin to understand the concepts of latitude and longitude.

Longitude.
In the diagram below,
C represents the centre of the Earth.
N & S represent the North and South poles respectively.

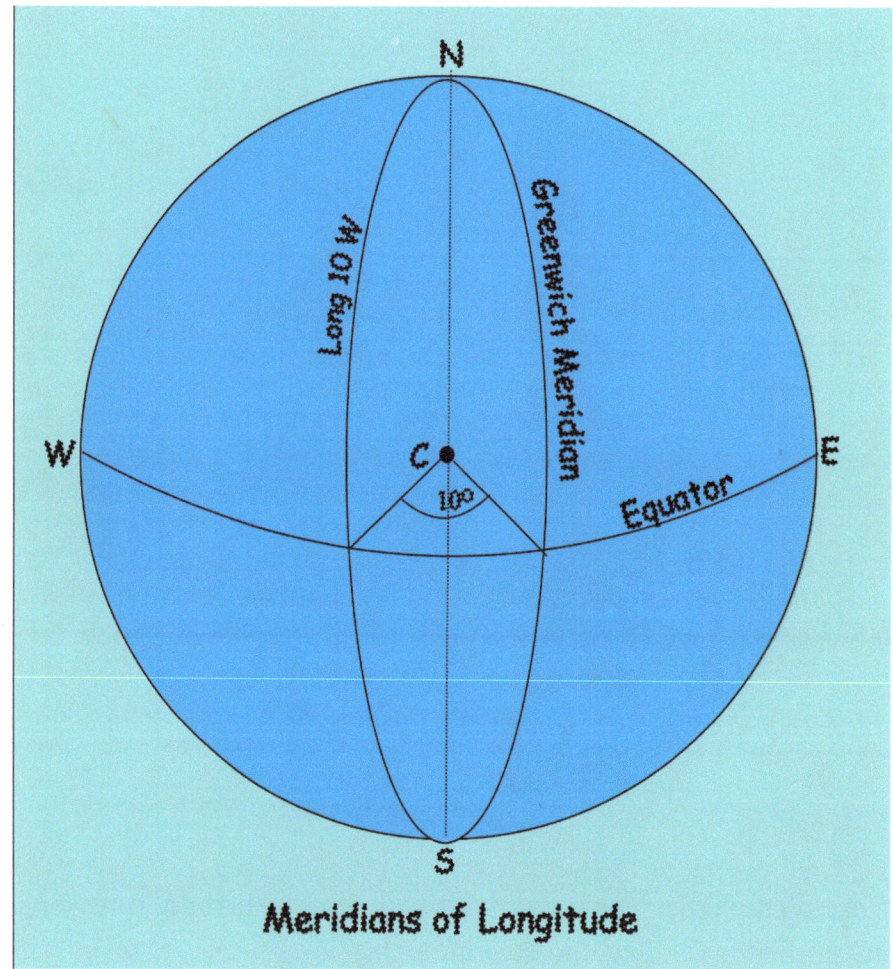

Meridians of Longitude

The great circles that pass through N and S are called **meridians of longitude**. The meridian that passes through Greenwich is used as the base meridian (0°) and all other meridians are described by their angular distance east or west of 0° along the Equator from 1° to 180°. The diagram shows the meridians of longitude 0° (Greenwich Meridian) and longitude 10° West.

The Equator. The Equator is an imaginary line around the Earth forming a great circle that is equidistant from the north and south poles and as such, it forms the boundary between the northern and southern hemispheres.

Latitude. A section of the Earth's surface made by a plane parallel to the Equator is called a parallel of latitude and by the definitions established above is a small circle.

A parallel of latitude is expressed by its angular distance north or south of the Equator. As illustrated in the diagram below, all points along the parallel of Latitude 10° N have an angular distance of 10° North from the Equator.

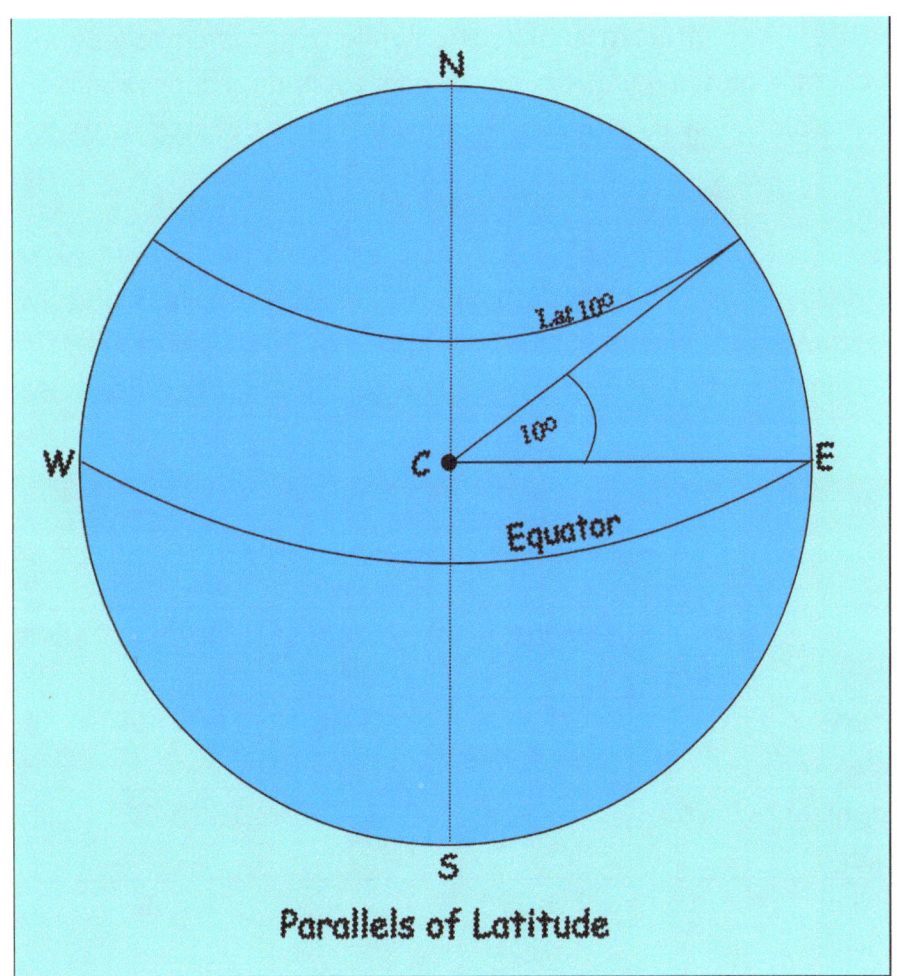

Parallels of Latitude

Units of Length.
The standard unit of measurement used for navigation at sea is the nautical mile. However, there are other units of length associated with the word mile and this can lead to confusion for navigators. The most common of these measurements are defined below:

Statute Mile. A mile most commonly refers to the statute mile of 5,280 feet (1,760 yards, or 1,609.344 meters). The use of the statute mile as a unit of measurement is largely confined to the United States and the United Kingdom; elsewhere, it has been replaced by the kilometer as a unit of measurement on land.

Geographical Mile. The international geographical mile (g.m.) is a unit of length determined by 1 minute of arc along the Earth's equator and is defined as 1855.32 metres.

Nautical Mile. The international nautical mile (n.m.) is closely related to the geographical mile and is a unit of length corresponding approximately to one minute of arc along any meridian of longitude. It is defined as exactly 1852 metres.

Kilometre. The kilometre (km) is a unit of length in the metric system and is equal to 1000 metres. As stated above, it is used to express distances between geographical places in most countries of the world except for the United Kingdom and the United States where the statute mile is used.

Measurement Conversion Table					
	Statute Miles	Geographical Miles	Nautical Miles	Metres	Kilometres
Statute Mile	-	0.867	0.869	1609.34	1.6
Geographical Mile	1.153	-	1.002	1855.32	1.86
Nautical Mile	1.15	0.998	-	1852	1.85
Kilometre	0.62	0.539	0.54	1000	-

The Circumference of the Earth. The Earth is not a true sphere and so there can be no single value for its circumference. The rapid rotation of the

Earth about its axis tends to flatten it into an **oblate spheroid** which simply means that the distance between the poles is shorter than the diameter of the equatorial axis giving it the appearance of a squashed ball. The result of this is that the circumference decreases as you move from the equator towards the poles.

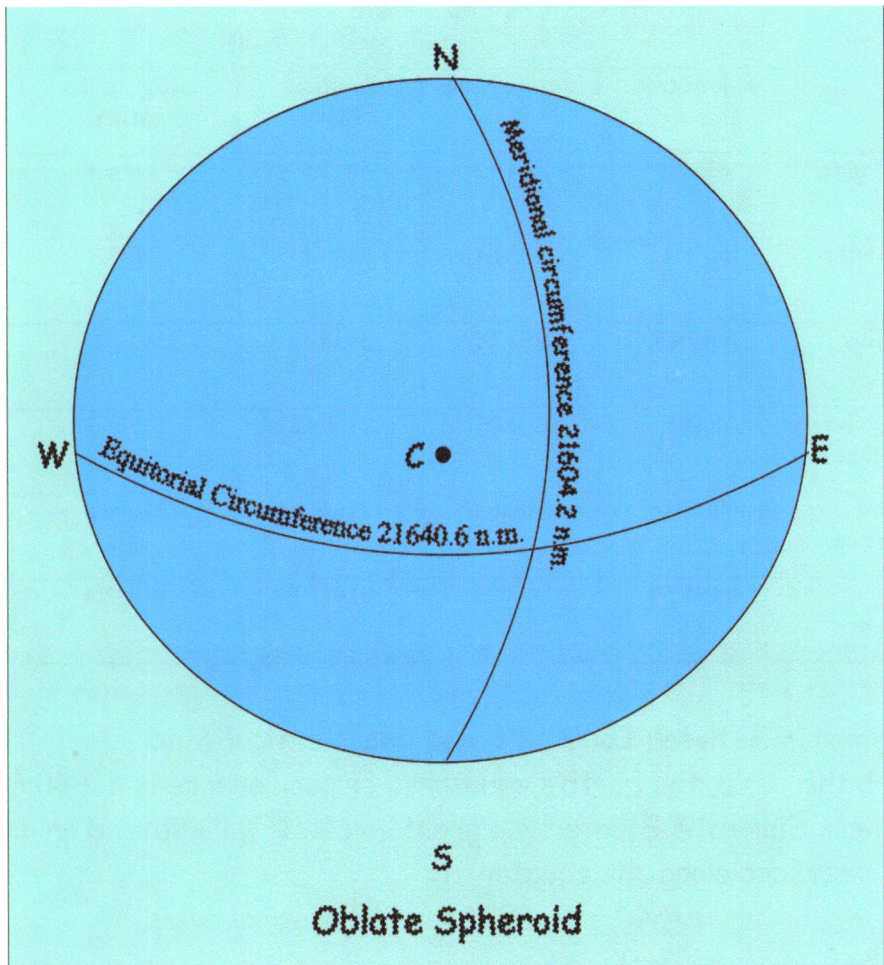

Oblate Spheroid

The three values of the Earth's circumference that we are most interested in are as follows:

Equatorial Circumference = 40075.16 Km.

Polar (or Meridional) Circumference = 40008 Km.

Mean circumference = 40041.58 Km.

For navigational purposes, we need to express measurements, sometimes in terms of nautical miles and sometimes in terms of geographical miles. We also need to understand various other dimensions of the Earth and be able to convert these from one unit of measurement to another.

The following table lists the most commonly used earth dimensions expressed in terms of different units of measurement.

Earth Dimensions Table				
	Kilometres	Statute Miles	Nautical Miles	Geographical Miles
Meridional (Polar) Radius	6356.8	3941.2	3432.67	3426.3
Equatorial Radius	6378.1	3954.4	3444.17	3437.8
Mean Radius	6367.45	3947.8	3438.42	3432.06
Meridional (Polar) Circumference	40008	24805	21604.2	21564.3
Equatorial Circumference	40075.16	24846.6	21640.6	21600.5
Mean Circumference	40041.58	24825.8	21622.5	21582.4

The Relationship Between Longitude and the Nautical Mile.
As shown in the table, the Earth's equatorial circumference is 21640.6 nautical miles. Since the Equator is a great circle, 1° will subtend an arc on the Earth's surface along the equator of:

$$\frac{21640.6}{360} = 60.113 \approx 60 \text{ nautical miles.}$$

There are 360 meridians of Longitude so it follows that, measuring from the Earth's centre; the angular distance between adjacent meridians at the Equator is 1°. Since 1° subtends an arc of 60 nautical miles, it also follows that the distance between adjacent meridians of longitude at the Equator is 60 n.m.

Returning to the longitude diagram below, the angular difference between longitude 10° West and the Greenwich Meridian is 10°; therefore, the distance between them at the Equator is 10 x 60 = 600 n.m.

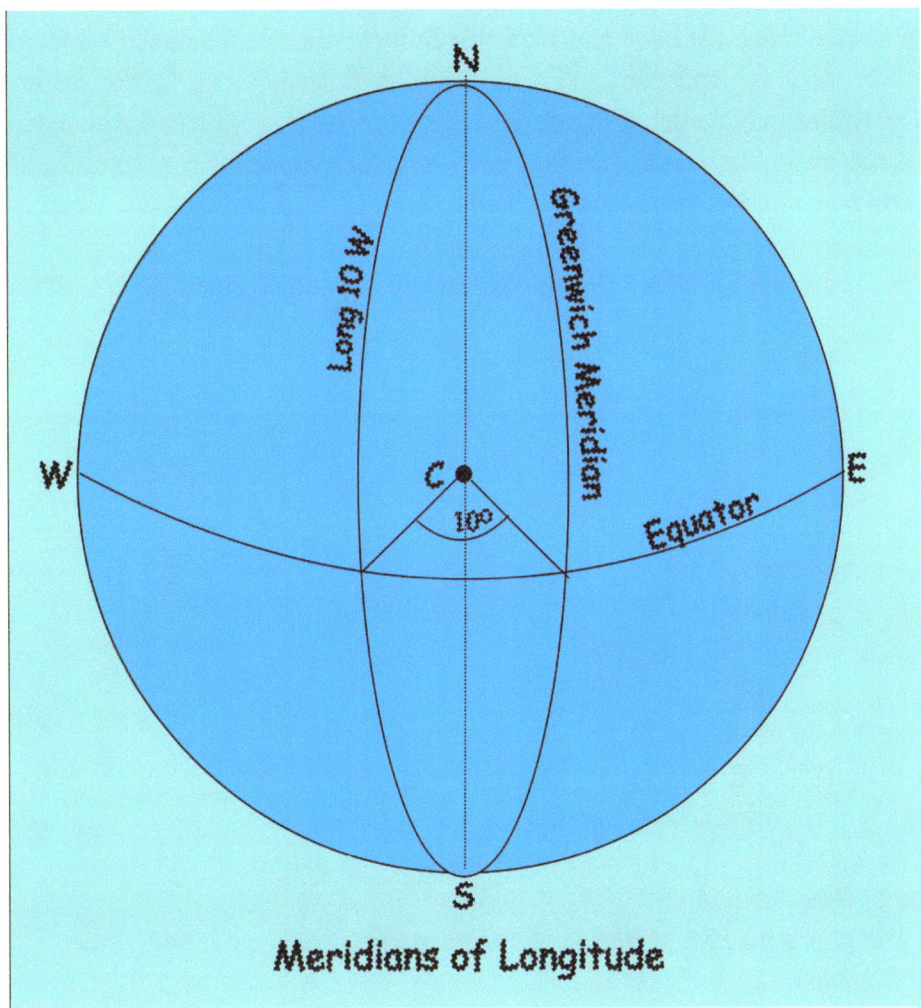

Meridians of Longitude

Time difference between meridians of longitude.

We know that the Earth revolves about its axis with respect to the Sun once every 24 hours. In other words, the Sun completes its apparent revolution of 360° in 24 hours. This means that the Sun crosses each of the 360 meridians of longitude once every 24 hours.

So, in 1 hour, the mean sun appears to move 15°,

in 4 minutes, it appears to move 1°,

in 1 minute it appears to move 15',

in 4 seconds it appears to move 1'.

From this, it becomes obvious that there is a direct relationship between arc and time.

Converting Arc to Time and Vice Versa.

From the above, the local hour angle of the Sun can be expressed in terms of both arc and time and it is important to be able to convert from one to the other. Conversion tables are available in publications such as the Nautical Almanac but if these are not available, it is quite easy and in fact more accurate, to make the conversions by using the following methods:

It is useful to remember the following when making conversions:

$$15° \leftrightarrow 1h$$
$$1° \leftrightarrow 4m$$
$$15' \leftrightarrow 1m$$
$$1' \leftrightarrow 4s$$

To Convert Arc Into Time.

Multiply by 4 and divide by 60

Example. Convert 65° 30' to time:

	h	m	s
4 x 65° ÷ 60 = 260° ÷ 60 =	4	20	00
4 x 30' ÷ 60 = 120' ÷ 60 =		2	00
∴ 65° 30' =	4	22	00

To Convert Time Into Arc.

Multiply the hours by 15 and divide the minutes and seconds by 4.

Example. Convert 12^h 32^m 15^s to arc:

$$12^h = 12 \times 15 = 180° \ 0' \ 0''$$
$$32^m = 32 \div 4 = 8° \ 0' \ 0''$$
$$15^s = 15 \div 4 = 0° \ 3' \ 45''$$
$$\therefore 12^h \ 32^m \ 15^s = 188° \ 3' \ 45''$$

The Relationship between Latitude and the Nautical Mile.

The Earth's meridional circumference is 40007.86 Km. which equates to 21604.2 nautical miles (n.m.). In other words, an angle of 360° at the Earth's centre subtends an arc of 21604.2 n.m. on the surface of a meridian of Longitude which is by definition a great circle.

From the above, it follows that:

1° measured along a meridian of longitude, will subtend an arc of:

$$\frac{21604.2}{360} = 60.012 \approx 60 \text{ n.m.}$$

and 1' will subtend an arc of:

$$\frac{60}{60} = 1 \text{ n.m}$$

The next diagram shows the parallel of latitude 1° North.
As calculated above, an angle of 1° at the Earth's centre will subtend an arc of 60 n.m. along a meridian of longitude. Therefore, any point on the parallel of latitude 1° North will have an angular distance of 60 n.m. north of the Equator.

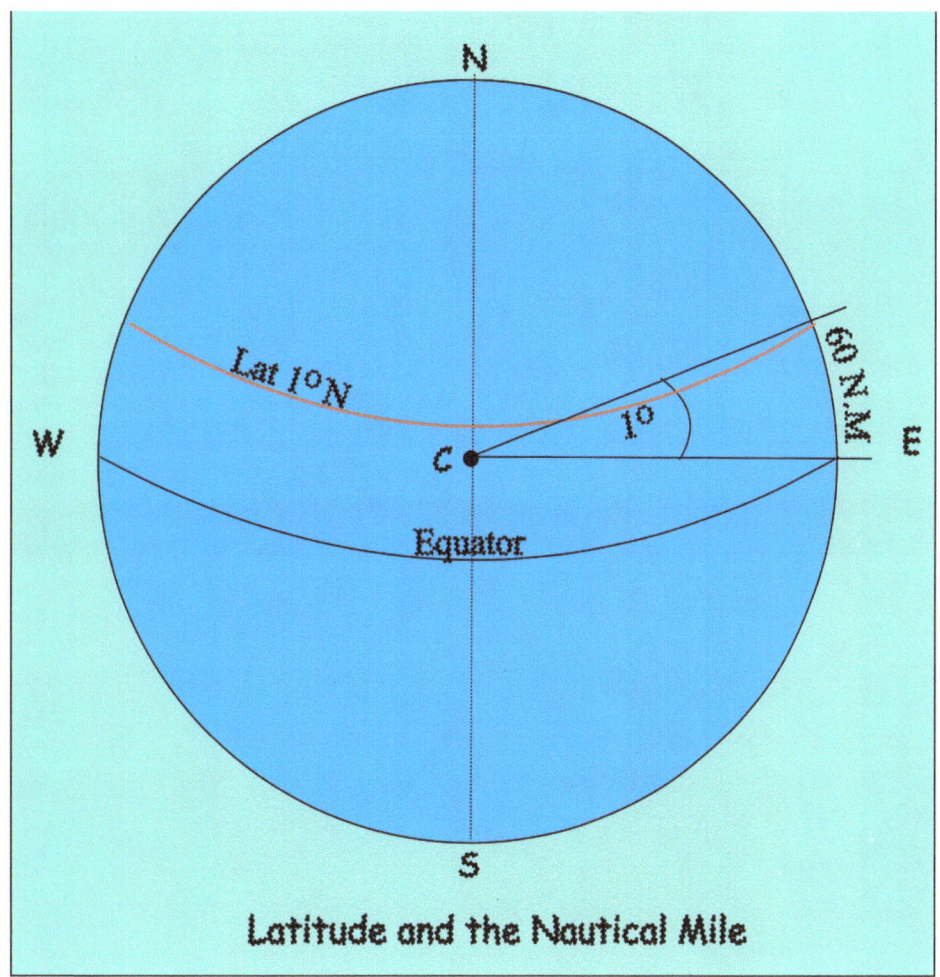

Latitude and the Nautical Mile

Distance Between Parallels of Latitude.

The angular distance between latitudes 62° N and 25° N is 37°. Therefore, the distance between these parallels, measuring due north or south along any meridian of longitude, will be: 37 x 60 = 2220 nautical miles as shown in the following diagram.

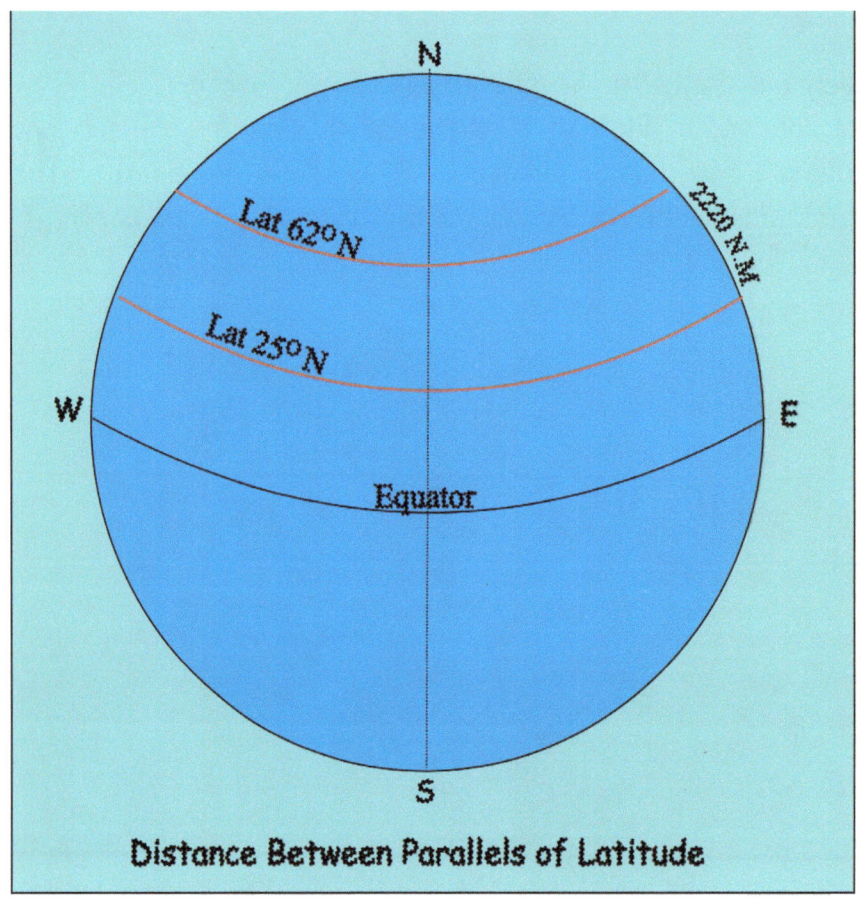

Distance Between Parallels of Latitude

Chapter 7
Time

Definitions of Time.

The apparent movement of the Sun provides the basis for our definitions of time.

The True Sun.

Life on Earth is governed by the movement of the True Sun, that is, the sun we see in the sky and not by a theoretical Mean Sun.

Apparent Solar Day.

The interval between two successive returns of the Sun to the local meridian is known as the apparent solar day.

Apparent Solar Time.

Also known as True Solar Time, apparent solar time is based on the apparent motion of the actual Sun, that is, it is based on the apparent solar day.

Apparent Noon is when the True Sun is on an observer's meridian of longitude. It is when the Sun reaches its greatest altitude above the observer's horizon. In other words it is when the Sun is at its **zenith**.

The Sun as a Time-Keeper.

Because the Earth's orbital motion is not uniform, there are corresponding variations in the apparent speed of the Sun along the ecliptic. For this reason, the hour angle of the True Sun does not increase at a uniform rate and therefore does not give an accurate measurement of time. To overcome this problem, the Mean Sun, as defined below, is used.

The Mean Sun.

The Mean Sun is an imaginary body which is assumed to move along the celestial equator at a uniform speed round the Earth and to complete one revolution in the time taken by the True Sun to complete one revolution of the ecliptic.

The Mean Solar Day is the time taken for the Mean Sun to make one complete circuit of the Earth. In other words, it is the time taken for the Mean Sun to transit all 360 meridians of longitude.

Mean Solar Time (usually abbreviated to **mean time)**. The time system based on the Mean Solar Day.

Mean Noon occurs when the meridian of the Mean Sun coincides with the meridian of a place. It is not the time when the True Sun reaches its zenith; that occurs at Apparent Noon.

The Civil Day is the day that is defined for use for human activity rather than for astronomical purposes. It begins at midnight when the Local Hour Angle of the Mean Sun is 12 hours or 180° and ends the following midnight. It is divided into two periods of 12 hours each. The first period consists of the 12 hours from midnight to noon and is denoted by the abbreviation a.m.

(ante meridian). The second period is the 12 hours from noon to midnight and is denoted by p.m. (post meridian).

The Astronomical Day consists of one period of 24 hours instead of two periods of 12 hours. This system is more convenient for tabulation purposes since times can be written as 4 figures and dispenses with the need for the abbreviations a.m. and p.m. For example, 11.45 a.m. is simply written as 1145 and 7.32 p.m. is written as 1932.

Local Hour Angle (LHA). As explained in chapter 2, the Local Hour Angle of a celestial body is the angle between the meridian of the observer and the meridian of the body's geographical position. LHA is measured westwards from the observer's meridian and can be expressed as angular distance or as time.

Local Mean Time (LMT) is the local hour angle of the Mean Sun measured westwards from the meridian of a certain place and expressed in terms of time instead of arc.
In the diagram which follows, imagine we are looking down on the North Pole from space.

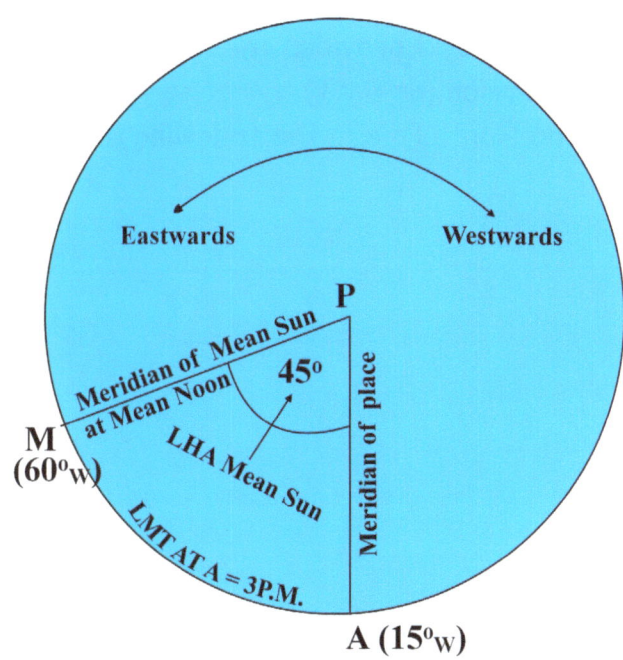

Local Mean Time

Point A is a place on the Earth's surface and AP represents part of the meridian of longitude of that place.

MP is part of the meridian of longitude on which, for a brief instant, the Mean Sun lies and M is a point on the Earth's surface which lies on that meridian.

Point P is the North Pole

Angle APM is the Local Hour Angle of the Mean Sun.

Suppose the meridian of the Mean Sun is 60°W and the longitude of point A is 15°W, then the LHA of the Mean Sun will be 45°. Since the Mean Sun moves 15° westwards in 1 hour, the time difference between point A and point M will be 3 hours. Because point A is 45° to the east of the Mean Sun's meridian, it must be 3 hours after Mean Noon and so the local mean time at point A will be 3 hours after noon or 3p.m.

Note. The difference between the LHA of the Mean Sun and the LHA of the True Sun is that the former is measured from the Mean Sun's meridian and the latter is measured from the meridian of the GP of the True Sun.

Greenwich Mean Time (GMT) is the local mean time anywhere on the meridian of Greenwich. In other words it is the Local Hour Angle of the Mean Sun on the meridian of Greenwich.

Since the Greenwich meridian is used as the base meridian from which the longitude of all places on Earth are identified, it provides the link between the LMT of a place and the LMT at Greenwich (or GMT).

Imagine that we are looking down on the North Pole in the following diagram.

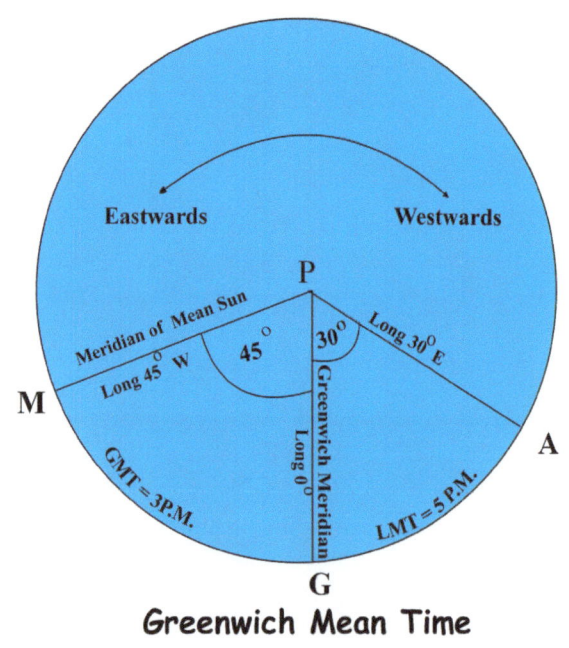

Greenwich Mean Time

P represents the North Pole

G represents the position of Greenwich on the Earth's surface.

GP represents part of the Greenwich Meridian (0°).

MP represents part of the meridian of longitude 45°W and M is a point on that meridian.

AP represents part of the meridian of longitude 30° E and A is a point on that meridian.

The meridian of the Mean Sun, for a very brief instant, coincides with the meridian 45°W and so, at that instant, the Local Mean Time at point M is noon.

At the same instant, the Local Hour Angle of the Mean Sun at Greenwich is 45°. Therefore, the LMT at Greenwich must be 3p.m. since the time difference for 45° is 3 hours and Greenwich is to the East of M.

It follows that the Greenwich Mean Time must also be 3p.m. (since GMT is equal to the LMT at Greenwich).

The LMT at point A must be 2 hours after GMT (since the time difference for 30° is 2 hours and A is to the East of Greenwich).

Therefore, the LMT at point A must be 5p.m.

The following **Aide Memoire** is useful when trying to remember whether to add or subtract the time difference:

Long West, GMT Best.

Long East, GMT Least.

Examples.
1. When it is noon at Greenwich, the LMT at a place on longitude 15°W will be 11 a.m. (since the time difference for 15° is 1 hour and the longitude is west).
2. If Greenwich Mean Time is 4 p.m., the LMT at a place 45°E will be 7 p.m.
3. If the LMT at longitude 75°W is 0130, what is GMT? The time difference for 75° is 5 hours and since the longitude is West, GMT must be +5 hours. Therefore, GMT is 0630.
4. If the time is 2130 GMT, what is the LMT at longitude 150° 15'E.? The time difference is 10 hours, 1 minute. Since the longitude is East,

LMT must be greater than GMT. Therefore, LMT is [(2130 + 1001) – 2400] = 0731 the next day.

The Equation of Time.

Although the imaginary Mean Sun gives us an accurate measurement of time, it presents the navigator with a problem. When fixing a vessel's position by the midday sun, the altitude of the true sun is measured at the moment that it reaches its zenith. The problem is that the true sun reaches its zenith at apparent noon in apparent solar time whereas the navigator measures the time of the altitude reading using a deck watch that keeps mean solar time. To enable us to connect mean solar time with apparent solar time, we have the Equation of Time which is explained below.

The True Sun moves with varying speed along the path of the ecliptic while the Mean Sun moves with constant speed along the path of the celestial equator. Because of this, their hour angles do not keep in step with each other and therefore we must be able to connect mean solar time with apparent solar time. The equation of time is designed to provide this connection and is defined as the excess of mean time over apparent time; that is:

Equation of Time = Local Hour Angle of Mean Sun - Local Hour Angle of True Sun
This can be abbreviated to **LHAMS - LHATS**

At certain times of the year, LHATS will be greater than LHAMS and at other times, it will be less with the values ranging from +15 to -15.
From 15th June to 31st August and from 25th December to 14th April, LHAMS is less than LHATS so the values of the equation of time are negative.
From 15th April to 14th June and from 1st September to 24th December, LHATS is less than LHAMS and so the values are positive.

The above is demonstrated in the following diagrams.

Notes pertaining to the diagrams:
1. Local Hour Angle (LHA) is measured westwards from the meridian of the observer).
2. TS: True Sun.   LHATS: LHA of True Sun.

3. MS: Mean Sun.   LHAMS: LHA of Mean Sun.
4. Z: Zenith of observer.
5.   Symbol for First Point of Aries :   ♈

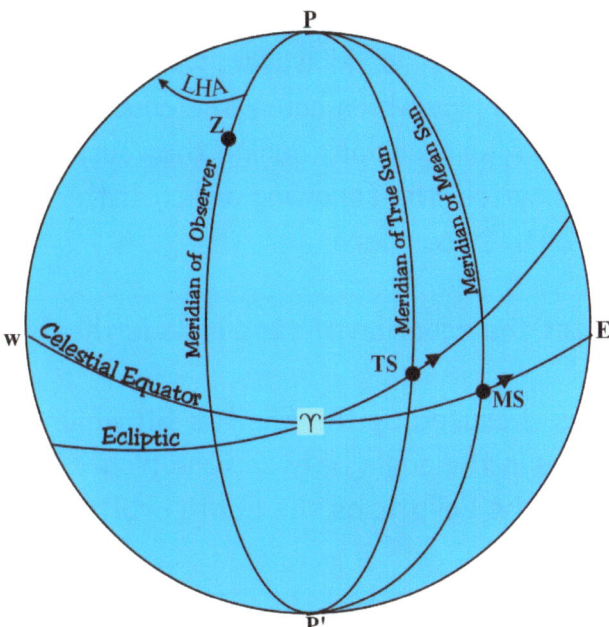

Equation of Time Negative
(LHAMS < LHATS)

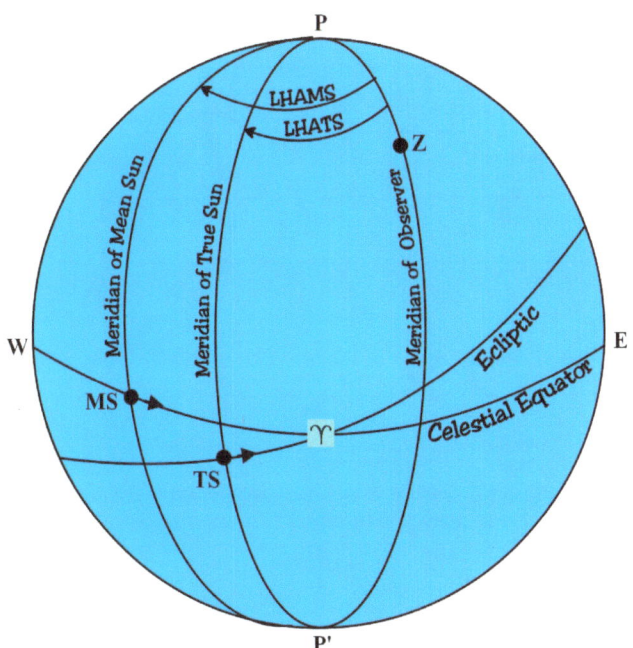

Equation of Time Positive
(LHAMS > LHATS)

♈ The First Point of Aries is the point at which the True Sun crosses the celestial equator as it moves northwards along the ecliptic. This event occurs at the Vernal Equinox and when it does, the equation of time becomes zero. We know that the True Sun crosses the celestial equator again at the Autumnal Equinox and it would seem that these are the only two occasions when the equation of time becomes zero. However, things are never that simple; other factors come into play which cause the equation of time to become zero and change sign on four occasions during the course of a year. These occasions fall on approximately the following dates: 15th April, 14th June, 1st September and 24th December.

In brief, the other factors that affect the equation of time include the following:

      a. **Obliquity.** This is caused by the Earth's tilt which leads to changes in the angle between the planes of the equator and the ecliptic as the Earth orbits around the Sun.

      b. **Eccentricity.** This is caused by the attraction of the Sun, Moon and planets on the Earth which results in the unequal motion of the Earth as it travels around its elliptical orbit.

The chart below shows that the equation of time becomes zero and changes sign on four occasions during the course of a year.

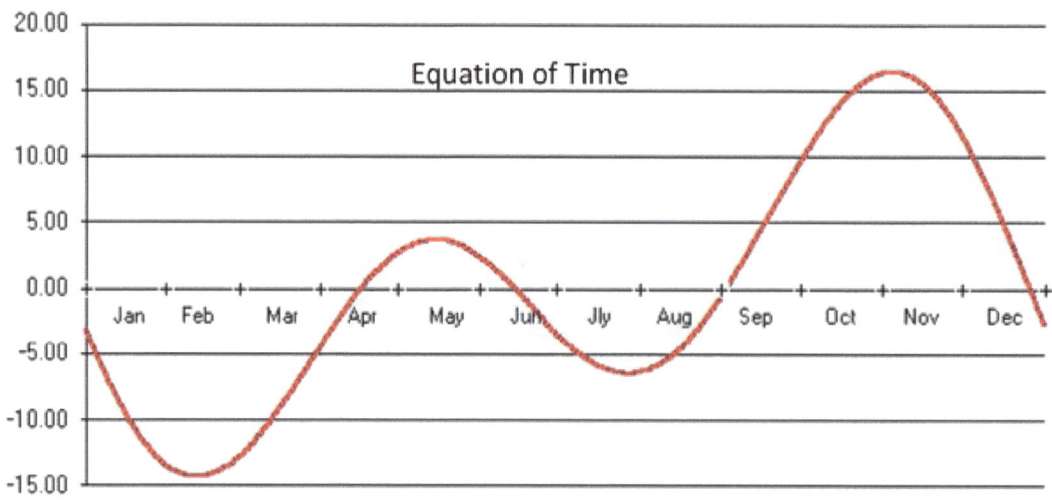

Summary.

- The equation of time can be either positive or negative depending on the time of the year.
- The maximum value is ± 16.4 but on average the values range from +15 to -15 mins.
- The values are positive from 15th April to 14th June and from 1st September to 24th December.
- The values are negative from 15th June to 31st August and from 25th December to 14th April.

Meridian Passage. Meridian passage is the instant that the True Sun crosses the local meridian of longitude and is the instant that it reaches its zenith. To simplify matters for the navigator, the Equation of Time for 00h (lower meridian) and 12h (upper meridian) for each day is printed at the foot of the Nautical Almanac daily page as shown in this extract. The Local Mean Time of the Sun's Meridian Passage is shown in the column to the right of the EOT. (This is the apparent time of Mer Pas adjusted for EOT to give the LMT and rounded up to the nearest minute).

	SUN		
Day	Eqn. of Time		Mer.
	00ʰ	12ʰ	Pass.
d	m s	m s	h m
21	01 42	01 49	12 02
22	01 55	02 02	12 02
23	02 08	02 15	12 02

If the mean time of Mer. Pas is shown to be greater than 1200 then the EOT must be negative, indicating that apparent time is slow compared to mean time. Conversely, if the mean time of Mer. pas. is less than 1200 then EOT is positive, indicating that apparent time is ahead of mean time.

Calculating Longitude by the Time of Meridian Passage. To calculate longitude we simply find the difference between the Local Mean Time (LMT) of Mer. Pas. and the Greenwich Mean Time (GMT) of our observation of Mer. Pas. then, by converting the time difference to arc we are able to find the difference in degrees of longitude.

Converting Arc to Time and Vice Versa. As previously discussed, due to the Earth's rotation, the Sun appears to move through 360° of longitude in 1 day, 15° in

1 hour and 15' in 1 minute of time and therefore, the local hour angle of the Sun can be expressed in terms of both arc and time.

It is useful to remember the following when making conversions:

$$15° \leftrightarrow 1h$$
$$1° \leftrightarrow 4m$$
$$15' \leftrightarrow 1m$$
$$1' \leftrightarrow 4s$$

Conversion tables are available in publications such as the Nautical Almanac but if these are not available, it is quite easy and in fact more accurate, to make the conversions by using the following methods:

To Convert Arc Into Time.

Multiply by 4 and divide by 60

Example. Convert 65° 30' to time:

	h	m	s
4 x 65° ÷ 60 = 260° ÷ 60 =	4	20	00
4 x 30' ÷ 60 = 120' ÷ 60 =		2	00
∴ 65° 30' =	4	22	00

To Convert Time Into Arc.

Multiply the hours by 15 and divide the minutes and seconds by 4.

Example. Convert 12^h 32^m 15^s to arc:

$$12^h = 12 \times 15 = 180° \ 0' \ 0''$$
$$32^m = 32 \div 4 = 8° \ 0' \ 0''$$
$$15^s = 15 \div 4 = 0° \ 3' \ 45''$$
$$\therefore 12^h \ 32^m \ 15^s = 188° \ 3' \ 45''$$

Let's try an example:

Date: 22 June. Zone Time: 1140 (+4). DR Pos: 32° 30'N. 61° 55'W

At meridian passage, the deck watch time was 16^h 08^m 25.1^s and the Deck Watch Error was -05.0^s. The daily page for that date shows that the Eqn. of Time is 02m 02^s and that Mer. Pas. is 1202 indicating that EOT is negative so apparent solar time is slow compared to mean solar time.

Calculate Longitude.

- **Calculate time difference.**

Deck Watch Time:	16^h 08^m 25.1^s
DWE:	-05.0^s
GMT:	16^h 08^m 20.1^s
LMT Mer.Pas:	12^h 02^m 00^s
Time Diff:	04^h 06^m 20.1^s

(Longitude West, GMT Best)

- **Convert Time to Arc**

$$4^h = 4 \times 15 = 60° \ 00' \ 00''$$

06^m = 6 ÷ 4 =   1° 30' 00"
20.1^s = 20.1 ÷ 4 = <u>0° 05' 01".5</u>
                  = 61° 35' 01".5

 Therefore Long = 61° 35' 01".5 W

Universal Time (UT). The term Universal Time was adopted internationally in 1928 as a more precise term than Greenwich Mean Time, because GMT can refer to either an astronomical day or a civil day.

Coordinated Universal Time , abbreviated as **UTC**, is the primary time standard by which the world regulates clocks and time. It is, within about 1 second of mean solar time at 0° longitude.   It is one of several closely related successors to **Greenwich Mean Time** (GMT). For most purposes, UTC is considered to be interchangeable with GMT, but GMT is no longer precisely defined by the scientific community.

However, the term Greenwich Mean Time persists in common usage to this day and is generally considered to be synonymous with the term Universal Time. It should be noted that the Nautical Almanac and other tables of astronomical data usually refer to UT instead of GMT.

Standard Time. For each place on Earth to keep its own local mean time would obviously cause a great deal of confusion and difficulty. For this reason, Standard Time is introduced so that places in the same locality can keep the same time. The time chosen is usually based on a convenient meridian running through the area and the meridian chosen usually differs from the Greenwich meridian by a whole number of hours. Some large countries, such as the U.S.A. have several different standard times.

Zone Time (ZT). It would be impossible for a ship at sea to keep to the time of its longitude because (unless it is travelling due north or south) the longitude will be constantly changing. For this reason, the sea areas of the Earth are divided into time zones which are 15° (or 1 hour) apart.   The central meridian of each zone is an exact number of hours distant from the Greenwich meridian so that zone time differs from GMT by multiples of 1 hour. The time kept in each zone is the time of its central meridian and is plus or minus GMT depending on the zone's position east or west of Greenwich.

Time Zone System. The Earth is divided into 24 zones of 15° of longitude each, with the centre of the system being the Greenwich meridian.

Therefore, the centre zone (zone 0) lies between 7.5° W. and 7.5° E. The zones lying to the West of Greenwich are numbered from +1 to +12 and those to the East from -1 to 12. The 12th zone is divided by the meridian 180° which is known as the International Date Line. The two halves of the 12th zone are marked + or - depending on which side of the date line they lie.

The following table shows the centre meridian of each of the 24 time zones. It should be remembered that each zone is 15° wide and extends 7.5° either side of its centre meridian. Thus, since the centre meridian of zone +4 is 60° W, its limits are 52.5° W to 67.5°W.

0	+1	+2	+3	+4	+5	+6	+7	+8	+9	+10	+11	12
0	15w	30w	45w	60w	75w	90w	105w	120w	135w	150w	165w	180

0	-1	-2	-3	-4	-5	-6	-7	-8	-9	-10	-11	12
0	15E	30E	45E	60E	75E	90E	105E	120E	135E	150E	165E	180

To Convert Zone Time to GMT. The time kept in zone 0 is GMT. To convert any other zone time to GMT, simply apply the sign and number of the zone to the zone time. For example, if it is 1800 in zone +5, GMT will be 2300.

The Greenwich Date (G.D.) is obtained by converting the zone time and date to GMT
Examples:
ZT 0230(-8), 15 Apr = GD 1830, 14 Apr.
ZT 1915(+10), 24 Dec = GD 0515, 25 Dec.

Practical Units of Time. The following units of time are defined in terms of the True Sun and although they are not all used in everyday life, they are important for navigational purposes and therefore need to be defined clearly.

The Year is the time taken for the Earth to complete one orbit of the Sun.

The Day is the time taken for the Earth to complete one revolution about its own axis. In other words, it is the time taken for the True Sun to make an apparent transit of all 360 meridians of longitude.

The Hour. The day is divided into 24 hours and in 1 hour, the Sun will make an apparent transit of 15° of longitude.

The Minute. The hour is divided into 60 minutes and in 1 minute of time, the Sun will transit 15 minutes (15') of longitude or, in other words, a $\frac{1}{4}$ of 1° of longitude.

The Second. The minute is divided into 60 seconds and in 1 second of time, the Sun will transit 15 seconds (15") of longitude or, in other words, a $\frac{1}{4}$ of 1' of longitude.

Chapter 8
Using Celestial Bodies For Navigation

We can define a celestial body's position in relation to our own celestial meridian and celestial horizon, in other words, in terms of its azimuth and altitude from our position. Conversely, we can establish our position in relation to the azimuth and altitude of a celestial body, the geographical position (GP) of which is known. Therefore, measuring azimuth and altitude are fundamental steps in the process of calculating a vessel's position.

An outline of the theory of astro navigation was given in chapter 1 of this book; however, a summary of the most important aspects of the subject is given below.

The altitude is the angle between the celestial horizon and the direction of the celestial body.

Azimuth is a specific type of bearing which measures the direction of an object in relation to true north, in the horizontal plane, clockwise from 0° to 360°.

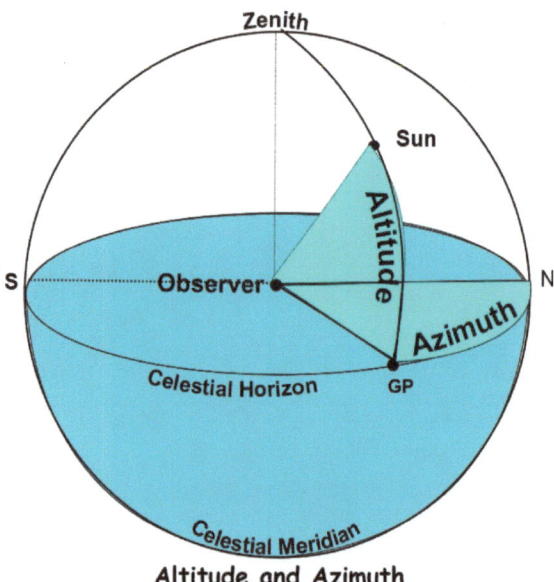

Altitude and Azimuth

Azimuth Angle. In astro navigation, when we calculate the azimuth of a celestial body, the result of the calculation is expressed as an azimuth angle.

Azimuth angle is measured from 0° to 180° either westwards or eastwards from either north or south. If the observer is in the northern hemisphere, the azimuth is measured from north and if in the southern hemisphere, it is measured from south.

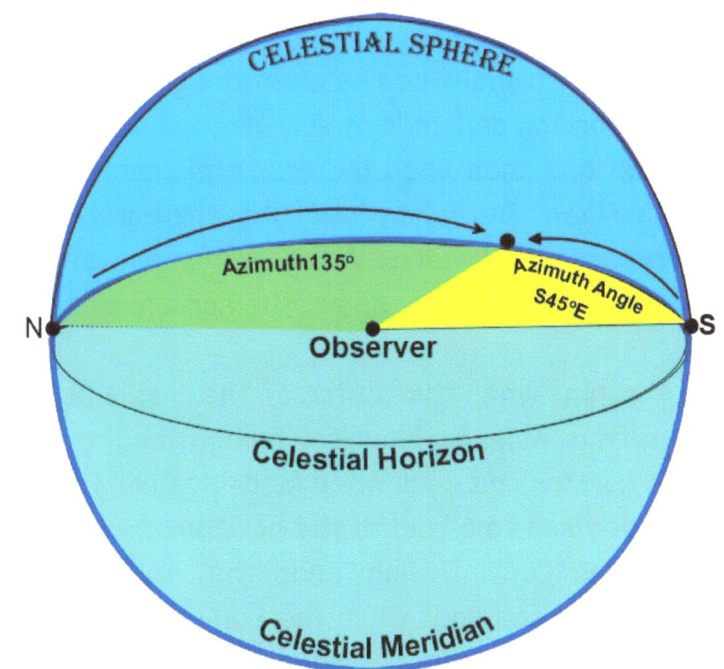

Azimuth and Azimuth Angle (Observer in Southern Hemisphere)

For example, if the true azimuth of an object is 225°, the azimuth angle for an observer in the northern hemisphere will be N135°W but for an observer in the southern hemisphere, it will be S045°W.

Celestial Bodies Used For Astro Navigation.
As well as the Sun and Moon which we can obviously use, certain stars and planets can also be used as follows.
Planets. Only those planets that are sufficiently prominent to be observed with an ordinary sextant are considered to be 'navigational planets'. These are: Venus, Mars, Jupiter and Saturn.
Stars. Only those stars that are sufficiently prominent in terms of magnitude and position are considered to be 'navigational stars'. There are 59 navigational stars and these are listed in the Nautical Almanac. A list of

navigational stars can also be found in chapter 10 of this book together with their magnitude, sidereal hour angle (SHA) and declination.

Twilight Time.   In general, the optimum conditions for taking observations of stars and planets occur during the times of civil twilight and nautical twilight when it is likely to be light enough for the horizon to be seen yet dark enough for those celestial bodies to be visible.

Civil Twilight. Morning civil twilight begins when the geometric centre of the Sun is 6° below the horizon and ends at sunrise. Evening Civil Twilight begins at sunset and ends when the geometric centre of the Sun reaches 6° below the horizon. During civil twilight, the horizon is clearly visible and the brightest stars as well as Venus, the brightest of the planets, can be seen as long as they are above the horizon at that time.

Nautical twilight is the time when the centre of the Sun is between 6° and 12° below the horizon. Thus, Morning Nautical Twilight begins when the Sun reaches 12° below the horizon and ends when Morning Civil Twilight begins, that is when the Sun has risen to 6° below the horizon. Evening Nautical Twilight begins when Evening Civil Twilight ends, that is when the Sun has sunk to 6° below the horizon and ends when the Sun is 12° below the horizon. Nautical twilight is so named because it is when navigators are able to take reliable sights of stars and planets using a visible horizon for reference. During Nautical Twilight, the horizon is visible and most of the navigational stars and all of the navigational planets can be seen as long as they are above the horizon at that time.
Venus is the third brightest celestial body in the sky and this planet can sometimes be observed later than morning nautical twilight or earlier than evening nautical twilight.

Positions.   In navigation, we speak about several different types of position and it would be sensible at this stage to clarify the differences between them:

Dead Reckoning (D.R.)   The term Dead Reckoning is used to cover all positions that are obtained from the course steered by the ship and her speed through the water since her last position and from no other factors. The D.R. position is therefore only an approximate position.

The D.R. position is shown on the chart as a small cross with the description and time marked alongside it thus:

$+$ DR 1100

Estimated Position (E.P.) The E.P. is derived from the D.R. position and is adjusted for the estimated effects of wind, current and tidal stream. However, it is still an approximate position because the exact influence on the ship's course and speed of those factors cannot be accurately assessed. The E.P. is shown on the chart as a dot surrounded by a small triangle, with the description and time alongside it, thus:

E.P. 1200

Assumed Position. When working with sight reduction tables, we use an 'assumed position' to help us to keep the number of interpolations to a minimum. We can do this because, when we calculate the intercept, we assume that the vessel is in the vicinity of the D.R. position which of course is not an accurate position. We can therefore, within reasonable limits, choose a position in the neighbourhood of the D.R. position that will not only save time and work but will also reduce the risk of making arithmetic errors.

Position Fix: The position of a vessel established by the intersection of two or more position lines is known as a position fix. Position lines may be obtained from a variety of sources such as visual bearings, electronic navigation aids, radar, and in the case of astro navigation, astronomical observations.

Observed Position: Where position lines are derived from astronomical observations, the resultant fix is known as an observed position and is marked on the chart with a small circle with the description and time alongside it, thus:

O Obs. 1240

Geographical Position. At a given time, any celestial body is located directly over one point on the Earth's surface. The latitude and longitude of that point is known as the celestial body's geographic position (GP).

Zenith. The Zenith is an imaginary point on the celestial sphere directly above the observer. It is the point where a straight line drawn from the geocentric centre of the Earth, through the observer's position and intersects with the celestial sphere.

The Zenith Distance. The Zenith Distance is the angular distance from the observer's zenith to the celestial body measured from the Earth's centre.

Greenwich Hour Angle (GHA). The angle between two meridians of Longitude can be expressed as an hour angle. The hour angle between the Greenwich Meridian and the meridian of a celestial body is known as the Greenwich Hour Angle.

Local Hour Angle (LHA). The LHA is the angle between the meridian of the observer and the meridian of the geographical position of the celestial body (GP).

How To Calculate The LHA Of A Star. The LHA of stars is not listed in the Nautical Almanac so we must calculate this ourselves. The following method can be used.

From the nautical almanac daily pages, find the Greenwich Hour Angle (GHA) of Aries at the planned time of the observation.

From the 'Index to Selected Stars' in the Nautical Almanac, find the Sidereal Hour Angle (SHA) of the chosen star.

Calculate your estimated longitude at the planned time of the observation.

Combine the SHA, GHA Aries and the estimated longitude to find the approximate LHA.

Example. Calculate the LHA of star Aldebaran in the following scenario: (Note. GHA and Longitude given to nearest whole degree to simplify explanation).

Details of Star Aldebaran at time of nautical twilight:
SHA: 291
Declination: 19°N
GHA Aries: 275°
Estimated longitude of observer: 135°W

To Calculate LHA of Star Aldebaran
SHA: 291

```
GHA Aries      275
               566
Long           -135  (subtract long if west, add long if east)
               431
               -360 (subtract 360 if LHA is greater than 360)
LHA    =        71°
```

How to Calculate the LHA Of the Sun, the Moon and the Navigational Planets.

Unlike the stars, the positions of the Sun, the Moon and the planets in the celestial sphere are not fixed and for this reason they are listed in the daily pages of the Nautical Almanac by their GHA and declination instead of by their SHA. The method used to calculate the LHA of a planet is the same as that for a star except that because the GHA is given in the Nautical Almanac, we do not need to calculate it.

Example. Calculate the LHA of planet Mars in the following scenario:

Details of planet Mars at time of nautical twilight:

GHA: 054°

Declination: 20°S

Estimated longitude of observer: 135°W

To Calculate LHA of Planet Mars:
```
GHA           054
Long          -135 (subtract long if west, add long if east)
LHA   =       -81
              +360 (add 360 if LHA is less than 0)
               279°
```

Declination. The declination of a celestial body is its angular distance North or South of the Celestial Equator. The declinations of the stars change very slowly and can be considered to be almost constant for up to a month at a time.

The declination of the Sun changes relatively fast from 23.4° North to 23.4° South and back again during the course of a year.

The Moon's declination is more difficult to predict because the rate of change is even more rapid than that of the Sun and the pattern of the changes is less uniform.

Like the Sun and the Moon, the declinations of the planets also change rapidly in comparison with the stars.

Declination can be summarised as the celestial equivalent of Latitude and for practical reasons, we treat it as the angular distance of a celestial body North or South of the Equator.

The Astronomical Position Line. A position line is a line drawn on a nautical chart along which a vessel's position is known to lie. When a position line is derived from an observation of a celestial body, it is known as an astronomical position line.

The Theory Behind Astro Navigation.

So far, we have considered azimuth and altitude from a position on the surface of the Earth. To fully understand how these phenomena relate to the LHA and declination of a celestial body and hence, how they help us to establish our position, we need to consider them in relation to the celestial sphere.

Consider the diagram below:

The celestial sphere is drawn in the plane of the observer's meridian with the observer's zenith (Z) at the top.

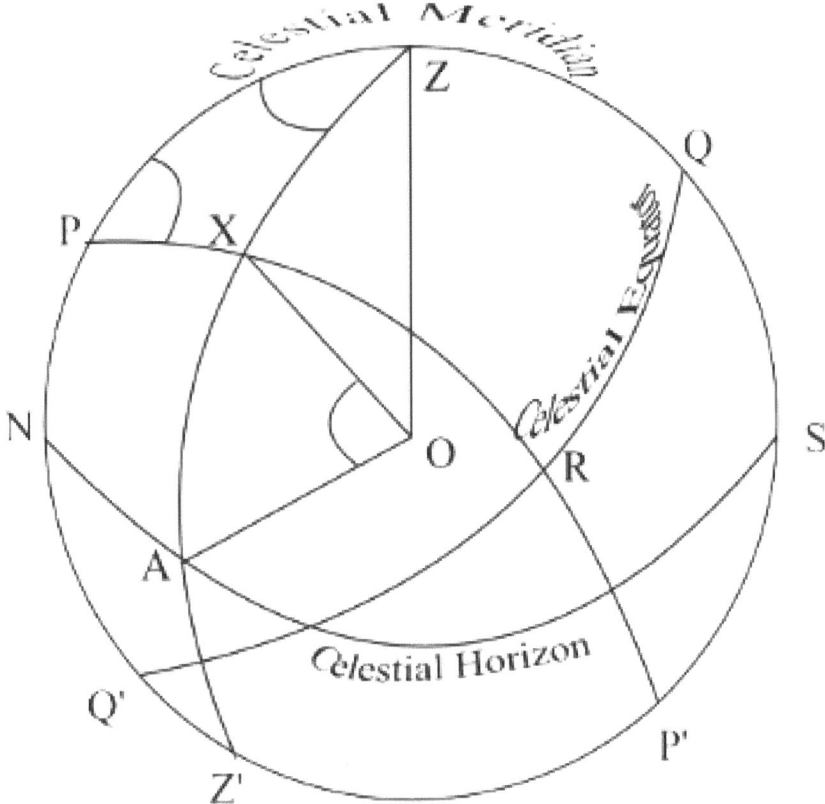

Point O represents both the observer and the Earth.

The arc PZQSP' represents the observer's celestial meridian.

The arc NAS is the celestial horizon and QRQ' represents the celestial equator.

ZXAZ' is a vertical circle running through the position of the celestial body (X). (A vertical circle is a great circle that passes through the observer's zenith and is perpendicular to the celestial horizon).

The Azimuth Angle is the angle PZX (that is, the angle between the observer's celestial meridian and the vertical circle through the celestial body).

The Altitude is the angle AOX (that is the angle from the celestial horizon to the celestial body measured along the vertical circle).

The Zenith. Point Z in the diagram represents the observer's zenith which is an imaginary point on the celestial sphere directly above the observer. It is the point where a straight line drawn from the geocentric centre of the Earth, through the observer's position and onwards, intersects with the celestial sphere.

The Zenith Distance. The Zenith Distance is the angular distance from the zenith to the celestial body measured from the Earth's centre. In the diagram above, it is the angular distance ZX which is subtended by the angle XOZ.

Relationship between Altitude and Zenith Distance

Since the celestial meridian is a vertical circle and is therefore, perpendicular to the celestial horizon, it follows that angle AOZ is a right angle and angles AOX and XOZ are complementary angles. From this we can deduce that:

> **Zenith Distance = 90° – Altitude**

and **Altitude = 90° – Zenith Distance**

Calculating the Zenith Distance.

Consider the next diagram.

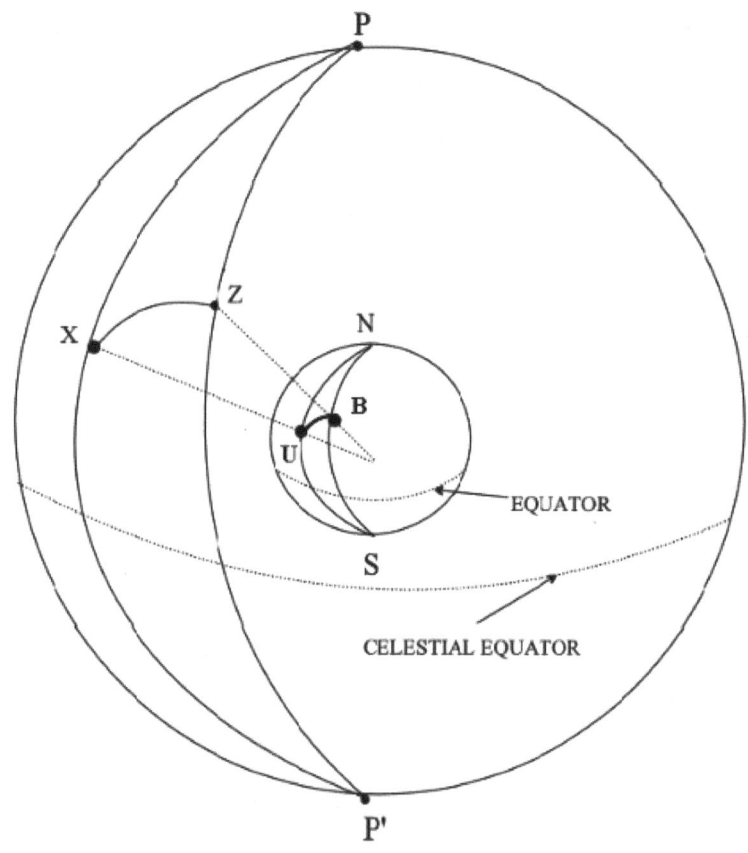

The diagram shows that the angular distance BU on the Earth's surface is equal to the angular distance ZX in the spherical triangle PZX.

In the diagram below,

X   represents the position of the Sun on the celestial sphere,

Z   represents a point on the sphere which coincides with the zenith of the DR position (B),

P   represents the projection of the North Pole onto the celestial sphere,

PX   = NU = (90° - the declination of the Sun),

PZ   = NB = (90° - the latitude of the DR position),

ZX   = BU = (90° - the altitude of the Sun).

We can see that the triangle NBU on the Earth's surface can be solved, in effect, by solving the triangle PZX in the celestial sphere.

(Although the Sun was used as the 'celestial body' for this explanation, the Moon, stars and planets could also be used).

Local Hour Angle (LHA)

In the PZX triangle diagram, LHA is the angle ZPX; that is the angle between the observer's celestial meridian and the meridian of the celestial body.

Relationship between LHA and Azimuth

Consider the next diagram.

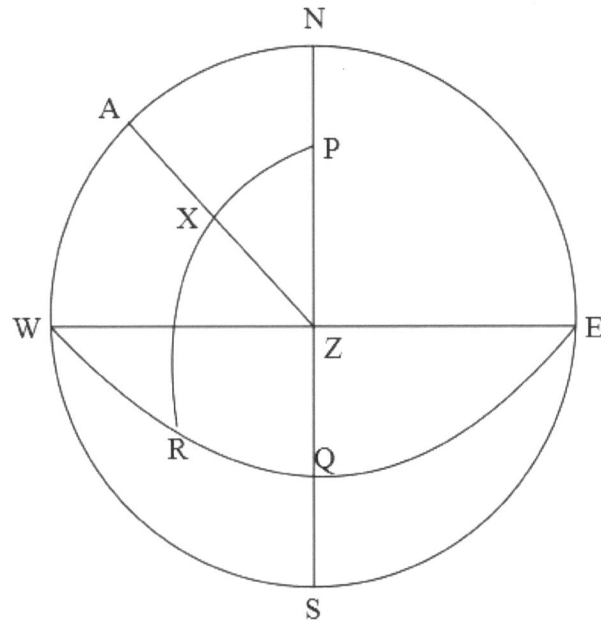

This diagram is drawn in the plane of the celestial horizon. Imagine that you are looking down on the celestial sphere from a position directly above the observer's zenith which is in the centre of the circle.

The circle WANESW represents the celestial horizon.

NZS represents the observer's celestial meridian.

WQE represents the celestial equator,

P is the celestial pole,

X is the position of the celestial body,

PXR represents part of the meridian of the celestial body which cuts the Equator at R.

ZPX is the LHA.

PZX is the Azimuth.

When the LHA (ZPX) is less than 180°, the celestial body lies to the west of the observer's meridian and when the LHA is greater than 180° it lies to the east. (Remember LHA is measured westwards from the observer's meridian from 0° to 360°).

It follows that if the celestial body is to the west of the observer's meridian, the azimuth must be west and when to the east, the azimuth must be east.

So we have the rule:

LHA 0° to 180° = Azimuth West

LHA 180° to 360° = Azimuth East

Summary Of The Discussions Above. The relationships discussed above illustrate the importance of azimuth and altitude in position finding at sea. The theory of astro navigation depends on the ability to solve the spherical triangle PZX and the azimuth and altitude give us the essential data we need to do this. With this data we are able to find the LHA, declination and zenith distance of a celestial body and armed with this information, we are able to establish our position on the Earth's surface. The process of calculating an astronomical position line from the altitude and azimuth is known as sight reduction and this is defined below.

Sight Reduction. This is the process of reducing the data gathered from an observation of a celestial body down to the information needed to establish an astronomical position line. The two essential items of data that are needed to begin the process of sight reduction are the azimuth and the altitude of the celestial body in question. Methods of sight reduction usually fall under two categories, tabular and formula. Tabular methods such as Rapid Sight Reduction involve interpolating large tables of data by entering latitude, declination and LHA to extract altitude and azimuth. Formula methods involve mathematically calculating the altitude and azimuth from the same input data.

(Note. The terms Astro Navigation and Celestial Navigation are considered to be synonymous).

Brief Outline of the Method of Calculating an Astronomical Position Line.
(Although the Sun is used as the 'celestial body' for this demonstration, the Moon, stars and planets could also be used).

Suppose we are in a yacht and we measure the altitude of the Sun which we find to be 35°; what does this tell us? All that we know is that the yacht lies somewhere on the circumference of a circle centered at the geographical position of the Sun. Such a circle is known as a **'position circle'** since the yacht is known to lie somewhere on its circumference.

The diagram below shows that, at any point on the circumference of the circle, the Sun's altitude will be 35° and that the distance of the yacht from the GP will be equal to the radius of the circle.

The problem is to establish at which precise point on the position circle the yacht lies.

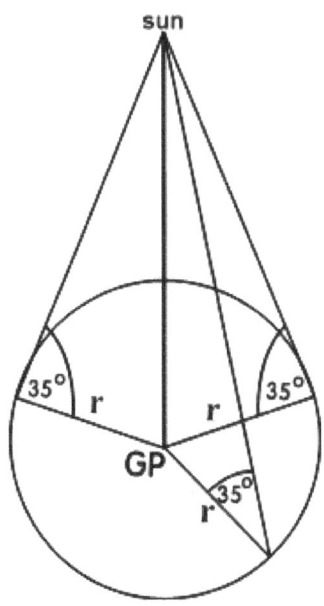

At first, it might seem that all we need to do is to observe the bearing of the Sun at the same time that we measure its altitude and then draw the line of bearing on the chart along with the position circle. In this way, it would seem that the yacht's true position would correspond to the intersection of these lines on the chart. However, there is a problem with this idea which makes it impracticable. Because of the great distance of the Sun from the Earth, the radius of the position circle will be very large (in the region of 3300 n.m.). A chart on which such a large circle could be drawn would require such a small scale that accurate position-fixing would be impossible.

However, we will know the yacht's D.R. position or E.P. which, although approximate, should be accurate to within a few minutes of latitude and longitude and this may give us another way of tackling the problem.

If we could calculate the altitude at the D.R. position for the time that the altitude is measured at the true position, we would then be able to calculate the zenith distances of the two positions. The difference between the two zenith distances would be equal to the distance from the DR position to the true position. The following few paragraphs explain how we can do this.

The Intercept. Our aim is to calculate the azimuth and altitude of a celestial body at the yacht's DR position for the time that we accurately measure the altitude at the true position. By finding the difference between the altitude at the DR position and the altitude at the true position and hence the difference between their zenith distances, we can calculate the intercept which is the distance from the DR position to the circumference of the position circle. The azimuth will give us a line of direction between the DR position and the geographical position of the celestial body.

Astronomical Position Line. We draw the position line at a point where the intercept intersects with the position circle. Since the circumference of a circle at any point is at right-angles to the radius at that point, no accuracy will be lost by drawing the position line as a tangent to the position circle. In this way, a navigator can draw the position line on the chart as a short, straight line which is perpendicular to the intercept as shown in the diagram below. Such a line is known as **an astronomical position line.**

Outline Method for Establishing An Astronomical Position Line.
The following short example, demonstrates the outline method for establishing an astronomical position line. However, to make the demonstration clear we will, for now, avoid 'clouding' our minds with all the mathematical calculations required. (A later example will show the calculations in full).

1. Measure The Altitude At The True Position And Then Calculate The Zenith Distance.

Suppose the altitude of the Sun, as measured at the true position of a ship, is $67°.972$. The zenith distance is equal to $90°$ – altitude so the calculation for finding the zenith distance from the altitude at the true position would be as follows:

Altitude of celestial body = $67°.972$

Zenith Distance = $90°$ – Altitude

= $90° - 67°.972 = 22°.028 = 1321'.68$

= 1321.68 nautical miles *(since 1 arc minute = 1 nautical mile on the surface of the Earth)*.

2. Calculate The Zenith Distance And Azimuth At The DR Position.

The method that we use to find the zenith distance and azimuth at the DR position contains a number of calculations; however, to keep this demonstration simple and easy to understand, we will assume that we have already calculated that the zenith distance at the DR position is $22°.142$ or $1328'.52$ which is equivalent to 1328.52 nautical miles on the surface of the Earth.

Assume also that we have calculated that the true azimuth at the DR position is $135°$.

3. Calculate the intercept.

Zenith distance of true position = 1321.68 n.m.

Zenith distance of DR position = 1328.52 n.m.

Therefore the Intercept = 1328.52 - 1321.68 = 6.84 n.m.

4. Plot the position line.
The intercept is drawn as a straight line from the DR position along the azimuth line of $135°$ as shown in the diagram below. The position line is drawn at right angles to the intercept at a point 6.84 n.m. from the DR position as shown in the diagram. (Note. Drawing not to scale).

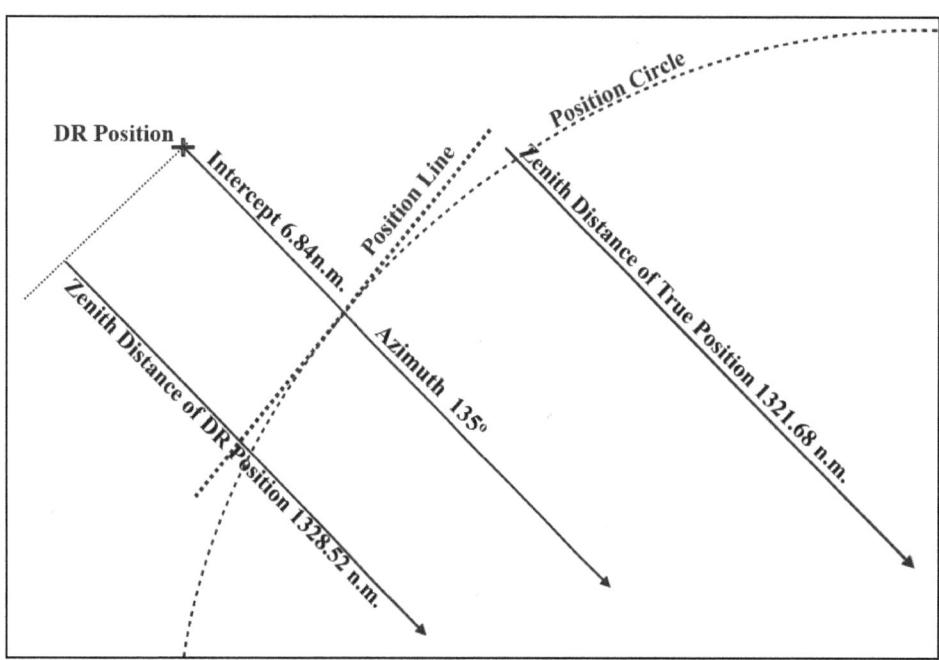

Note 1. We cannot assume that the yacht's position is at the exact point that the intercept meets the position circle because we are working from a DR position which is an approximate position. For this reason, we draw a short position line at right angles to the intercept; we can then assume that the correct position is at some point on this short line.

Note 2. A single position line does not constitute a position fix, all that it tells us is that a vessel's position lies somewhere along that line. For this reason, we require the intersection of two or more position lines in order to establish a fix.

Accuracy of Observed Positions Derived From Astronomical Position Lines. Observed positions derived from two or more astronomical observations cannot be entirely relied upon. The reasons for this are firstly, that the observations are not likely to be taken simultaneously since it is not possible to take sextant readings of more than one celestial body at the same instant. The faster a vessel travels, the greater the movement of the observer between the observations and the more significant this error becomes even when special methods of calculation are used. Secondly, observed altitudes are very seldom correct due to inherent errors in sextant readings and therefore, the resultant observed zenith distances will not be correct. For these reasons, the resultant position lines will be displaced. A summary of inherent sextant errors can be found at the end of this chapter.

Two Point Fix.   An error in just one position line is bound to lead to an error in the intersection point of the two lines thereby making the accuracy of the fix doubtful. Generally, a two-point fix should only be used to give an approximate position.

Three Point Fix. Position lines obtained from three astronomical observations are not likely to pass through a common point; instead, they will usually form a small triangle known as a 'cocked-hat'. Because the position within the cocked-hat is arrived at by guess-work, it is unlikely to be absolutely correct.

The diagram below illustrates how we can use the intersection of three astronomical position lines to establish a '3 point fix'. Note that the observed position is at the centre of a 'cocked-hat' triangle.

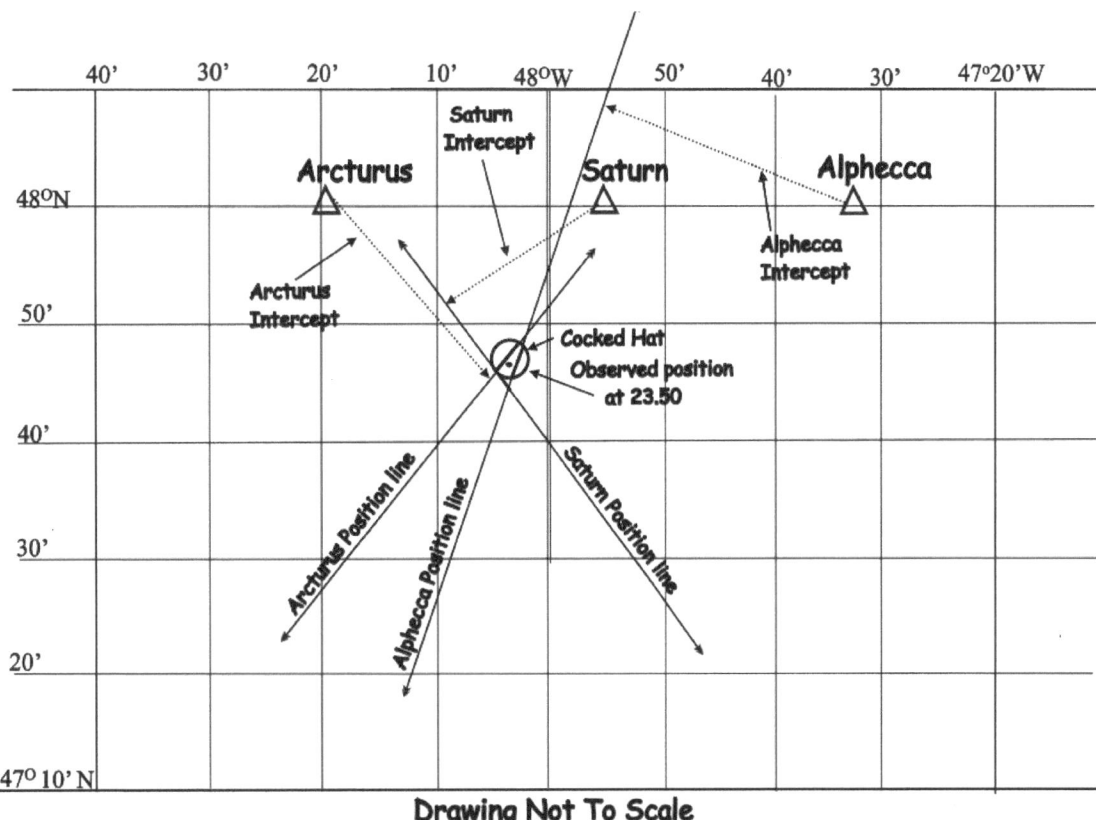

Drawing Not To Scale

Fixes from Sightings of Stars and Planets. There are 59 navigational stars and 4 navigational planets which we can use to achieve position fixes derived from three or more position lines but there are only two short periods during the day in which we can do this because we need it to be dark enough to see the bodies in the sky yet light enough to see the horizon. In other words, we are restricted to taking star and planet sights during the times of morning and evening nautical twilight.

Daylight Fixes. During the hours of daylight, we are mainly restricted to obtaining fixes from just one celestial body, the Sun. Sometimes, we can achieve a two point fix from the Sun and the Moon but mostly, all we can do is obtain two time-separated astronomical position lines using the Marcq St. Hilaire method to give us a two point fix.
(Note. The Marcq St. Hilaire method (more commonly known as 'Sun-Run-Sun') is fully explained in 'Astro Navigation Demystified).

The Moon. The moon can be used for position fixing when it is above the horizon as long as the horizon can be seen. There are three situations when this can occur: during nautical and civil twilight; at night as long as the moon creates sufficient light for the horizon to be seen and during daylight hours, particularly during early morning and late afternoon.

Rising and Setting Times.
The times of rising and setting of the Sun and Moon can be found in the daily pages of the Nautical Almanac so we can quickly see when they will be above the horizon. The Nautical Almanac also provides low to medium-precision positions of the Sun, and Moon and the navigational planets for the purpose of planning observations. 'The Nautical Almanac of the Stars' gives us the tools to calculate the rising and setting time of stars and there are many commercial software packages that also provide this facility.

Planning Observations of Stars. Because nautical twilight gives us only a short period of time to make observations, advanced planning is essential. We need to establish, in advance, what our estimated position at nautical twilight will be. We also need select the stars and planets that we intend to use for the fix (to be sure of obtaining three reliable sights for a three point fix, we should select at least four or five bodies). (Owners of this book have the advantage of being able to check which stars are 'in season' from chapter 5).

Having selected our stars and/or planets, the next step is to calculate what the altitude and azimuth of each of them will be at the DR position at the time of the planned observation. In this way, we will know what the approximate altitude and azimuth of each of the bodies will be when we come to make the sightings from the true position and this will cut the time we take to locate them in the sky.

There is also the added advantage that by making the calculations for the DR position in advance, most of the work required to calculate the intercept will have been completed by the time the observation is made. This will of course, speed up to process of calculating the position lines and the ultimate fix.

Using Spherical Trigonometry For Sight Reduction. With spherical trigonometry, we have the tools to quickly calculate the altitude and azimuth of selected celestial bodies at the estimated position without being encumbered by large tables of data.

Why Spherical Trigonometry? At first sight, the term 'spherical trigonometry' might seem quite daunting but with the knowledge of just two simple formulas and with a little practice of the methods demonstrated below, it will be found to be quick and easy to apply. The method is outlined in the next chapter and is comprehensively taught in my book 'Celestial Navigation - Theory and Practice'.

Why Calculate Azimuth? The true azimuth and the azimuth angle provide exactly the same directional information albeit in different formats. This begs the question: "why go to the trouble of calculating the azimuth angle and then converting it to the true azimuth when it is easier just to measure the true azimuth directly with a compass?" However, we calculate the azimuth by finding the angle PZX from the values of the sides PZ, PX and ZX in the spherical triangle ZPX and these values are derived from data relating to the DR position. (See the next chapter for an explanation of the ZPX triangle). If we measure the azimuth by compass, we can only do so from the true position. At the time of taking the altitude, we would not know where the true position is so our aim must be to find the direction of the true position from the DR position and we can only do this by calculating the azimuth angle at the DR position. There is also the point that, unless you are in the fortunate position of having a gyro compass, you must take magnetic compass readings and these have to be corrected for variation and deviation; so you might just as well calculate the true azimuth in the first place.

There is another important reason to be able to use spherical trigonometry for this task as the following statement in the 'International Maritime Organization Regulations' makes clear. "The provision of trigonometric tables onboard and regular practice with them by all officers and navigation related staff is compulsory". In the event of GPS and other electronic navigation systems failure, it would be irresponsible of a ship's master if his ship were to go dangerously off course simply because trigonometric calculations could not be made".

Example. Using Spherical Trigonometry To Calculate the altitude and azimuth of the star Alioth at the DR position at the planned time of observation using the data provided in the following scenario.

Scenario.
Estimated Position: Lat. 50°N   Long. 45°W
Data from Nautical Almanac re. Alioth:
SHA = 166°.
Declination = 56°N
GHA Aries = 300°

Step 1. Estimated position at planned time of observation:
Lat. 50°N   Long. 45°W

Step 2. From the Nautical Almanac, extract data for Alioth for planned time of observation as follows:
SHA = 166°.   Declination = 56°N.   GHA Aries = 300°

Step 3. Calculate LHA.
SHA Alioth   166
GHA Aries   <u>300</u>
                        466

Long   <u>-45</u>   (subtract if long is west, add if long is east)
              421   (subtract 360 if LHA is greater than 360)
            <u>-360</u>
LHA Alioth   <u>61°(W)</u>

Step 4. Calculate PZ, PX and ZPX (see next chapter for an explanation of these).
PZ = 90° - Lat = 90° - 50° = 40°
PX = 90° - Dec = 90° - 56° = 34°
ZPX = LHA = 61°(W)

Step 5. Calculate Zenith Distance (ZX).
Formula to Calculate ZX:
Cos (ZX) = [Cos(PZ) . Cos(PX)] + [Sin(PZ) . Sin(PX) . Cos(ZPX)]
(This formula is explained in the next chapter).
Enter data in the formula: (decimals to 3 places).
        = [Cos(40°) . Cos(34°)] + [Sin(40°) . Sin(34°) . Cos(61°)]
        = [0.766 x 0.829] + [0.643 x 0.559 x 0.485]
        = 0.635 + 0.174
Cos (ZX) = 0.809
∴ ZX   = Cos^{-1} (0.809)   =   36°

Step 6. Calculate Altitude.
Altitude = 90° - ZX
          = 90° - 36° = 54°.

Step 7. Calculate Azimuth.
Data previously calculated:
PX = 34°
PZ = 40°
ZX = 36°

Formula to Calculate Azimuth (PZX): (This formula is explained in the next chapter).
Cos PZX = <u>[Cos(PX) - (Cos(ZX) . Cos(PZ))]</u>
            [Sin(ZX) . Sin(PZ)]
Enter data into the formula: (decimals to 3 places).

$$\text{Cos PZX} = \frac{[\text{Cos}(34) - (\text{Cos}(36) \cdot \text{Cos}(40))]}{[\text{Sin}(36) \cdot \text{Sin}(40)]}$$

$$= \frac{[0.829 - (0.809 \times 0.766)]}{[0.588 \times 0.643]}$$

$$= \frac{0.829 - 0.620}{0.378}$$

$$= \frac{0.209}{0.378}$$

Cos(PZX) = 0.553

∴ PZX = $\text{Cos}^{-1}(0.553)$ = 56°.43 ≈ 56°

∴ Azimuth angle = N56°W (True Azimuth = 360° - 56° = 304°)

Summary: The altitude and azimuth of the star Alioth at the estimated position (EP or DR) at time of planned observation have been calculated in advance to be:

Altitude: 54°. Azimuth: 304°

This data serves two purposes:

Firstly it helps the navigator to quickly locate the position of the star at the true position by providing its approximate altitude and azimuth. Secondly, by calculating the altitude and azimuth from the EP or DR in advance, the navigator will have all the data necessary to quickly calculate the intercept.

Calculating the Intercept. (Full procedure explained in 'Astro Navigation Demystified')

Ho is observed altitude.

Hc is calculated altitude.

To calculate intercept: p = Ho - Hc

If p is positive, the intercept is from the DR position towards the azimuth.

If p is negative, the intercept is from the DR position away from the azimuth (i.e. towards the reciprocal).

Calculating the intercept for the example above.

Suppose the actual measurements of altitude and azimuth of Alioth at the true position at the time of observation were as follows:

Altitude: 53.785° Azimuth: 304°

We would calculate the intercept in the following way:

Calculated altitude (Hc) = 54°

Observed altitude (Ho) = 53.785°

p = Ho - Hc = 53.785° - 54° = -0.215° = -12.9' = -12.9 nautical miles
Therefore intercept = 12.9 n.m from 304° i.e. 12.9 n.m. towards 124° (reciprocal).

Note. 1 minute of arc at the Earth's centre will subtend a distance of 1 nautical mile at the Earth's surface. (See chapter 6,).

The Rapid Sight Reduction method is taught in my book 'Astro Navigation Demystified'.
The trigonometric method of sight reduction is taught in my book 'Celestial Navigation - Theory and Practice'.

Corrections to Sextant Readings. The following explanations of some of the corrections that have to be made to the sextant altitude in order to calculate the **True Altitude** have been copied from chapter 3 because they are also relevant to this chapter.
Corrections For Refraction. When a ray of light from a celestial body passes through the Earth's atmosphere, it becomes bent through refraction and this causes the apparent (observed) altitude to be greater than the true altitude. Since the sextant measures the apparent altitude, a correction for refraction must be applied to find the true altitude. Refraction is at its greatest when the altitude is small (i.e. when the celestial body is near the horizon) and becomes less as the altitude increases.
The effects of refraction are illustrated in the diagram below.

Effect of Refraction on Altitude

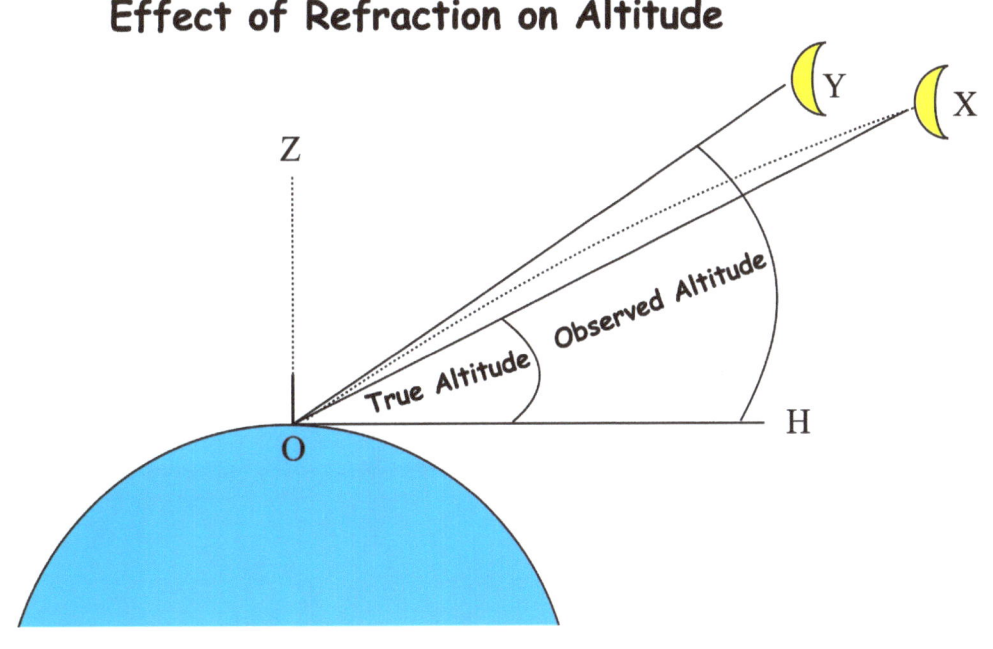

O is the observer's position and Z is the zenith at that point.

OH is the horizon

XOH is the true altitude of the celestial body from the observer's position. However, due to refraction, the celestial body appears to be at Y and so YOH becomes the observed altitude and a correction will have to be made to compensate for this.

Corrections For Parallax. We measure the altitude of a celestial body from our position in relation to our visible horizon; this is known as the **observed altitude**. However, when calculating the **true altitude**, measurements are made from the Earth's centre in relation to the celestial horizon. The displacement between the observed position of an object and the true position is known as **parallax**.

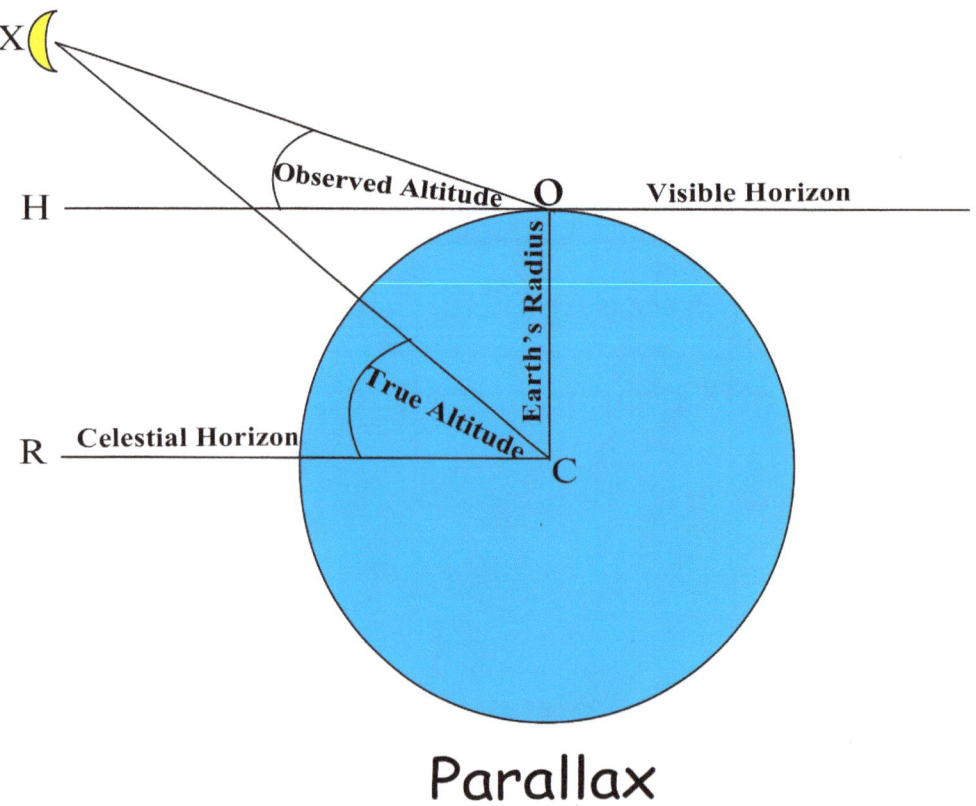

Parallax

Because of the vast distances of the stars and the planets from the Earth, we can assume that, in their cases, the celestial horizon and the visible horizon correspond with very little error. However, in the cases of the Sun and the Moon, which are relatively near, a correction called Parallax must be added.

Dip. A correction has to be made to allow for the height of the observer's eye above the horizon; this is known as Dip.
Consider the diagram below:
O is an observer's position on the Earth's surface and E is the position of his eye. We can see that, as the observer's height of eye is raised above sea level, his visible horizon 'dips' below the true horizon and so the altitude measured at E becomes greater than that measured at O.

Dip is the error caused by this difference and has to be subtracted from the sextant reading.

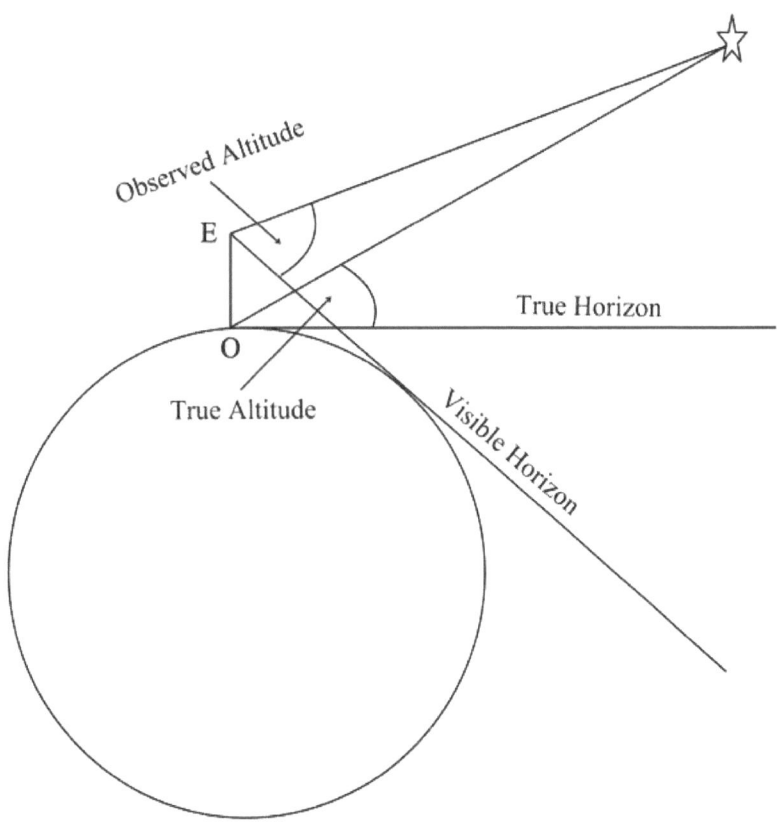

Chapter 9
Introduction to Spherical Trigonometry.

In practical astro navigation, we mostly rely on the use of tables of computed data and rote-learned procedures. We can operate quite efficiently in this way for most of the time; however, for those who wish to develop a thorough understanding of the subject, it is important to study the principles of spherical trigonometry which underpin astro navigation.

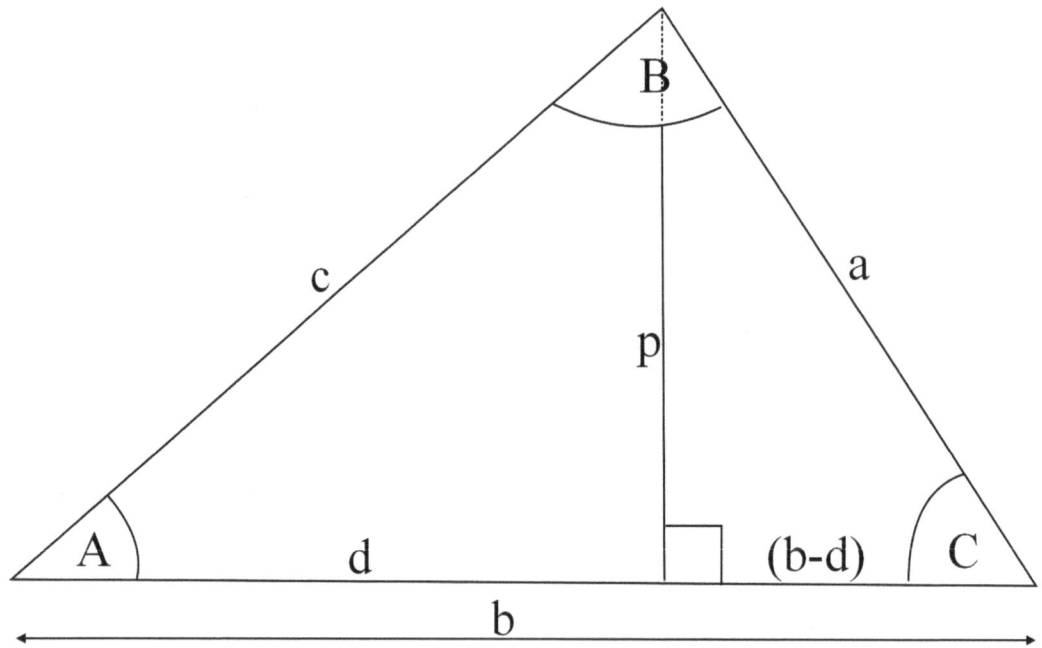

In the diagram above,

$\cos C = d/a$ and $d = a \cos C$

$P^2 = c^2 - (b-d)^2$ and $p^2 = a^2 - d^2$

$\Box\ a^2 - d^2 = c^2 - (b-d)^2$

$\Box\ a^2 - d^2 = c^2 - b^2 + 2bd - d^2$

$\Box\ a^2 = c^2 - b^2 + 2bd$

$\Box\ a^2 = c^2 - b^2 + 2b.a \cos C$ (since $d = a \cos C$)

$\Box\ c^2 = a^2 + b^2 - 2b.a \cos C$

This is the **cosine rule** for 'flat' triangles but does this rule also apply to spherical triangles?

The next diagram shows a spherical triangle ABC formed by the intersection of three circles with their common centre O at the centre of the sphere.

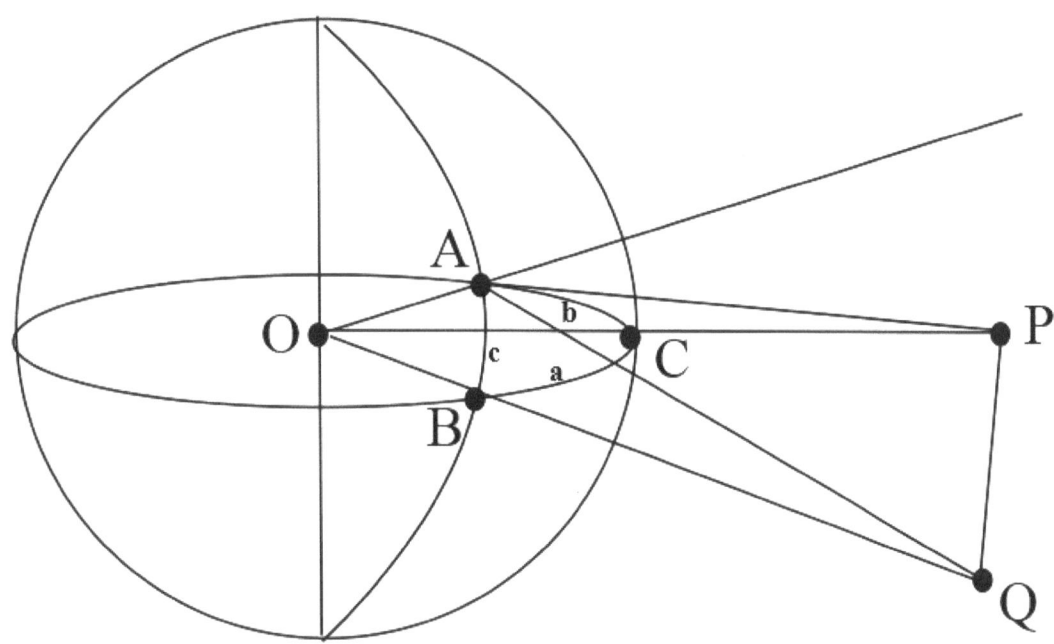

The edges of the 'flat' planes in which the sides a, b, & c lie, meet along OA, OQ and OP.

The edges AQ and AP just graze the great circles of the arcs c and b at A; that is, AQ is the tangent to c and AP is the tangent to b. So angles OAQ and OAP are really right angles although it is impossible to draw them as such in a spherical drawing.

The edges of the three planes in which a, b, & c lie form a flat (two-dimensional) triangle PAQ, of which the apical angle PAQ is equivalent to the angle A of the spherical triangle ABC.

Now, by the rule for 'flat' triangles, we have:
$$PQ^2 = PO^2 + QO^2 - 2PO \cdot QO \cdot Cos(a)$$
and $\quad PQ^2 = PA^2 + QA^2 - 2PA \cdot QA \cdot Cos(A)$
$\Rightarrow \quad [PO^2 - PA^2] + [QO^2 - QA^2] - [2PO.QO.Cos(a)] + [2PA \cdot QA \cdot Cos(A)] = 0$

We also have:

$PO^2 - PA^2 = AO^2$ and $QO^2 - QA^2 = AO^2$

$\Rightarrow$ $PO^2 - PA^2] + [QO^2 - QA^2] = 2AO^2$

Substituting $2AO^2$ in the equation gives us:

$2PO \cdot QO \cdot Cos(a) = 2AO^2 + 2PA \cdot QA \cdot Cos(A)$

Dividing through by $2PO \cdot QO$, we have:

$Cos(a) = \dfrac{AO}{PO} - \dfrac{AO}{QO} + \dfrac{PA}{PO} - \dfrac{QA}{QO} Cos(A)$

$= [Cos(POA) \cdot Cos(QOA)] + [Sin(POA) \cdot Sin(QOA) \cdot Cos(A)]$

$= [Cos(b) \cdot Cos(c)] + [Sin(b) \cdot Sin(c) \cdot Cos(A)]$

Hence, the formula for finding the third side (a) of a spherical triangle when the other two sides (b and c) are known together with the included angle (A) is: $Cos(a) = [Cos(b) \cdot Cos(c)] + [Sin(b) \cdot Sin(c) \cdot Cos(A)]$

(This is the cosine rule for spherical triangles).

PZX Triangle.

Consider the diagram below:

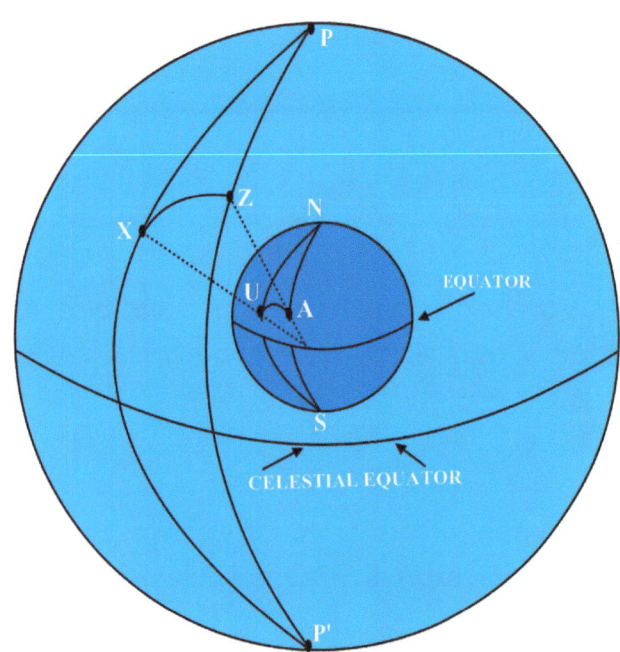

The PZX Triangle

Point B represents the position of an observer on the Earth's surface and Z represents the projection of B onto the celestial sphere. So Z is the point on the celestial sphere directly above the observer and is called the '**Zenith**

X represents the position of a **celestial body** and U represents its projection onto the surface of the Earth. So U is the point on the Earth's surface immediately below the celestial body and is called the '**Geographical Position' (GP).**

N is the Geographical North Pole; it is the projection the Celestial North Pole (P) onto the Earth's surface.

Similarly, S is the Geographical South Pole and is the projection of P'.

The arc ZX is the projection of the arc BU.

The Celestial Equator is the projection of the Earth's Equator,

The Celestial Meridian PZP' is the projection of the observer's meridian NBS,

The Celestial Meridian PXP' is the projection of the meridian NUS on which lies the geographical position.

As demonstrated in chapter 8, we apply the cosine rule in astro navigation when calculating the zenith distance which is the side ZX in the spherical triangle PZX as follows:

Cos ZX = [Cos PZ. Cos PX] + [Sin PZ. Sin PX. Cos ZPX]

Also as demonstrated in chapter 8, we can apply the cosine rule to the problem of calculating the azimuth angle as follows:

$$\text{Cos PZX} = \frac{[\text{Cos(PX)} - (\text{Cos(ZX)} . \text{Cos(PZ)})]}{[\text{Sin(ZX)} . \text{Sin(PZ)}]}$$

Chapter 10
Navigational Stars

The following is a list of the 59 navigational stars together with their magnitude, sidereal hour angle (SHA) and declination.

Note. Star Magnitude. It may seem illogical but the higher the value of its magnitude, the lower the brightness of the star. For example, the brightest star in the sky is Sirius with a magnitude of -1.47 while the dullest navigational star is Zubenelgenubi with a magnitude of 3.28. By comparison, the magnitude of the Sun is –27, a full moon is –13 and Venus, the brightest planet, is –5.

Name	Mag.	SHA	Dec.(°)
Acamar	3.2	315	S.40
Achernar	0.5	335	S.57
Acrux	1.3	173	S.63
Adhara	1.5	256	S.29
Aldebaran	0.9 (Var)	291	N.17
Alioth	1.8	167	N.56
Alkaid	1.9	153	N.49
Al Na'ir	1.7	28	S.47
Alnilam	1.7	276	S.1
Alphard	2.0	218	S.09
Alphecca	2.2	126	N.27
Alpheratz	2.1	358	N.29
Altair	0.8	62	N.9
Ankaa	2.4	353	S.42
Antares	1.09	113	S.26
Arcturus	0.0	146	N.19
Atria	1.9	108	S.69
Avior	2.4	234	S.59
Bellatrix	1.6	279	N.6
Betelgeus	0.58(Var)	271	N.7
Canopus	-0.72	264	S.53
Capella	0.71	281	N.46

Deneb	1.25	050	N.45
Denebola	2.14	183	N.15
Diphdar	2.04	349	S.18
Dubhe	1.87	194	N.62
Elnath	1.68	279	N.29
Eltanin	2.23	091	N.51
Enif	2.4	034	N.10
Fomalhaut	1.16	016	S.30
Gacrux	1.63	172	S.57
Gienah	2.8	176	S.17
Hadar	0.6	149	S.60
Hamal	2.00	328	N.23
Kaus Australis	1.8	084	S.34
Kochab	2.08	137	N.74
Markab	2.49	014	N.15
Menkar	2.5	315	N.04
Menkent	2.06	149	S.36
Miaplacidus	1.7	222	S.70
Mirfak	1.82	309	N.50
Nunki	2.06	076	S.26
Peacock	1.91	054	S.57
Polaris	2.01	319	N.89
Pollux	1.15	244	N.28
Procyon	0.34	245	N.05
Rasalhague	2.10	091	N.13
Regulus	1.35	208	N.12
Rigel	0.12	282	S.08
Rigil Kentaurus	-0.01	146	N.19
Sabik	2.43	103	S.16
Schedar	2.25	350	N.56
Shaula	1.62	097	S.37
Sirius	-1.47	259	S.17
Spica	1.04	159	S.11
Sulhail	2.23	223	S.43
Vega	0.03	081	N.39
Zubenelgenubi	3.28	138	S.16

Note. This list includes Polaris which is not listed as a navigation star in the nautical almanac issued by the United Kingdom Hydrographic Office. However, it has always played an important role in navigation; not only because it indicates the direction of north but also because it can be used for position fixing particularly in the polar-regions and for this reason, it is listed in the American Practical Navigator as a navigation star

INDEX

www.ingramcontent.com/pod-product-compliance
Lightning Source LLC
Chambersburg PA
CBHW050713180526
45159CB00003B/1020